CW01151505

CUSTOM PICK-UPS

CUSTOM PICK-UPS

Colin Burnham

OSPREY AUTOMOTIVE

Published in 1992 by Osprey Publishing Limited
59 Grosvenor Street, London W1X 9DA

© Colin Burnham 1992

All rights reserved. Apart from any fair dealing for the purpose of private study, research, criticism or review, as permitted under the Copyright, Designs and Patents Act, no part of this publication may be reproduced, stored in a retrieval system, or transmitted in any form or by any means, electronic, electrical, chemical, mechanical, optical, photocopying, recording or otherwise, without prior written permission. All enquiries should be addressed to the Publishers.

ISBN 1 85532264 4

Photography by Colin Burnham
Text by Richard Coney

Editor Shaun Barrington
Page design by Angela Posen
Printed in Hong Kong

About the author

Colin Burnham is a London-based photographer who has specialised in the American automotive field for many years. He has worked as a technical editor and features editor for Britain's *Street Machine* magazine and his pictures and words have appeared in numerous publications both in Europe and America. This is Burnham's sixth title in the Osprey Colour Series following *Air-cooled Volkswagens, Classic Volkswagens, California Classics, Bizarre Cars* and *Beverly Hills Cars*, but his first in collaboration with writer Richard Coney, a regular contributor to *Street Machine* and *Classic American*.

Front jacket

'Eliminator' is a Chevrolet-based monster from California. The costs involved in building and campaigning such a vehicle are sky-high (sic), and there are stringent safety precautions that builders must comply with before being allowed to compete at fairground tracks and indoor arenas. See 'Four-Wheel Jive' chapter for more monster-trucks

Rear jacket

There is no doubt about the function of this custom workhorse; Barnes VIP Wrecker Service will take your junker away in style

Half-title page

Klassic lines of the Rod and Custom *Magazine 'Dream Truck' live again after 30 years of limbo (see page 52)*

Title page

Strangely-flamed '55 Ford F-100 being pinstriped on Main Street, Hesperia, a desert town just east of Los Angeles

For a catalogue of all books published by Osprey Automotive please write to:
**The Marketing Department, Octopus Illustrated Books,
1st Floor, Michelin House, 81 Fulham Road, London SW3 6RB**

Contents

Introduction	6
Oldies but Goodies	8
Fabulous Fifties Rigs	50
Late-model Showcase	76
Four-wheel Jive	102

'Born-again' '39 Chevy has fresh paint, highly-polished brightwork and a hot-to-trot V8 beneath its louvred hood – what more does an old truck need?

Introduction

So why produce a book about pick-ups at all, let alone *custom* pick-ups? They're ugly, dirty, uncomfortable, limited in passenger space, usually knocked-about, mis-treated, abused and generally undesirable as a means of personal transport. There's certainly nothing intrinsically glamorous or classy about a pick-up truck.

Well, that may be the case in Britain and the rest of Europe, but in the USA the lowly pick-up is viewed in an entirely different light. Americans, on the whole, consider this breed of commercial workhorse to be socially acceptable in a way that Europeans, on the whole, have difficulty coming to terms with.

For years now Stateside enthusiasts have been fixing-up old trucks to better-than-new condition and driving them with pride. Even brand new ones, straight off the showroom floor, are customised – for the very reason that they *are* pick-ups. Today it is possible to order a domestic or imported truck from a dealer with more accessories, flashy paint and creature comforts than were available for passenger cars not so long ago.

It is often the case that you'll see more pick-ups on Main Street, strutting their stuff, than cars – and that's not including those loaded down with lumber or agricultural produce. These are the customised trucks used for commuting to work and cruising the night spots; vehicles that will never be used for hauling hay or driven across a muddy field or rutted track.

So how did this phenomenon come about, to turn a plain old down-at-heel utilitarean vehicle into a sporty, flashy, extravagantly-pampered and, in many cases, extremely expensive personalised pick-up?

First, it must be stated that the pick-up truck has always played an important role in the rather transient American way of life. It is a dual-purpose vehicle, capable of transporting both passengers and possessions from one dream to the next and equally practical during the interim periods of settlement. Consequently both privately and commercially owned pick-ups exist in large numbers all across the nation and it is this very omniprescence which makes them the perfect basis for customising. They are relatively simple machines; easy to work on and easy to understand. Bodylines are straight, upholstery is minimal, and engine swaps are a breeze. They are good vehicles for young hot rodders to learn on.

It all started during the post-World War II period when cars and trucks began to flood back onto a market that had been starved of new vehicles for too many years. While Mom and Dad revelled in the delights of a new sedan, Junior was given a free hand to fix-up the family's worn-out old pick-up for his own use. Once up and running, the truck, probably a 1930's Ford or Chevrolet, would have received an inordinate amount of tender loving care by its proud young owner, eventually incorporating many unique personal touches and improved performance.

As the Fifties progressed hot rodding became increasingly popular in the United States, such that many vehicles were taken off the road and completely stripped down prior to a lengthy transformation, or only used at weekends. Thus, in order to stand out from the crowd, the daily-driven pick-up received more and more attention until it eventually became an object of desire in its own right. The rest, as they say, is history...

Today, those car-crazy

teenagers of the Fifties and their offspring enjoy their pick-up trucks more than ever before. Pictured on the following pages is a fascinating cross-section of vehicles from all over America which illustrate the diversity of the trucking scene. From pre-war domestic products, rescued from farms and wrecking yards and totally rebuilt as street rods, through the semi-restored classics of the Fifties and radically modified later model trucks, to the very latest mini/monster-trucks. They're all here in their glorious colours. Enjoy!

Above
The cleaning team get to it at the first-ever Hot Rod Supernationals in Cleveland, Ohio. Where they'll all fit is another matter; there's not a lot of space in Henry Ford's '29 Model 'A' pick-up (remember the poor Joad family in The Grapes of Wrath*?)*

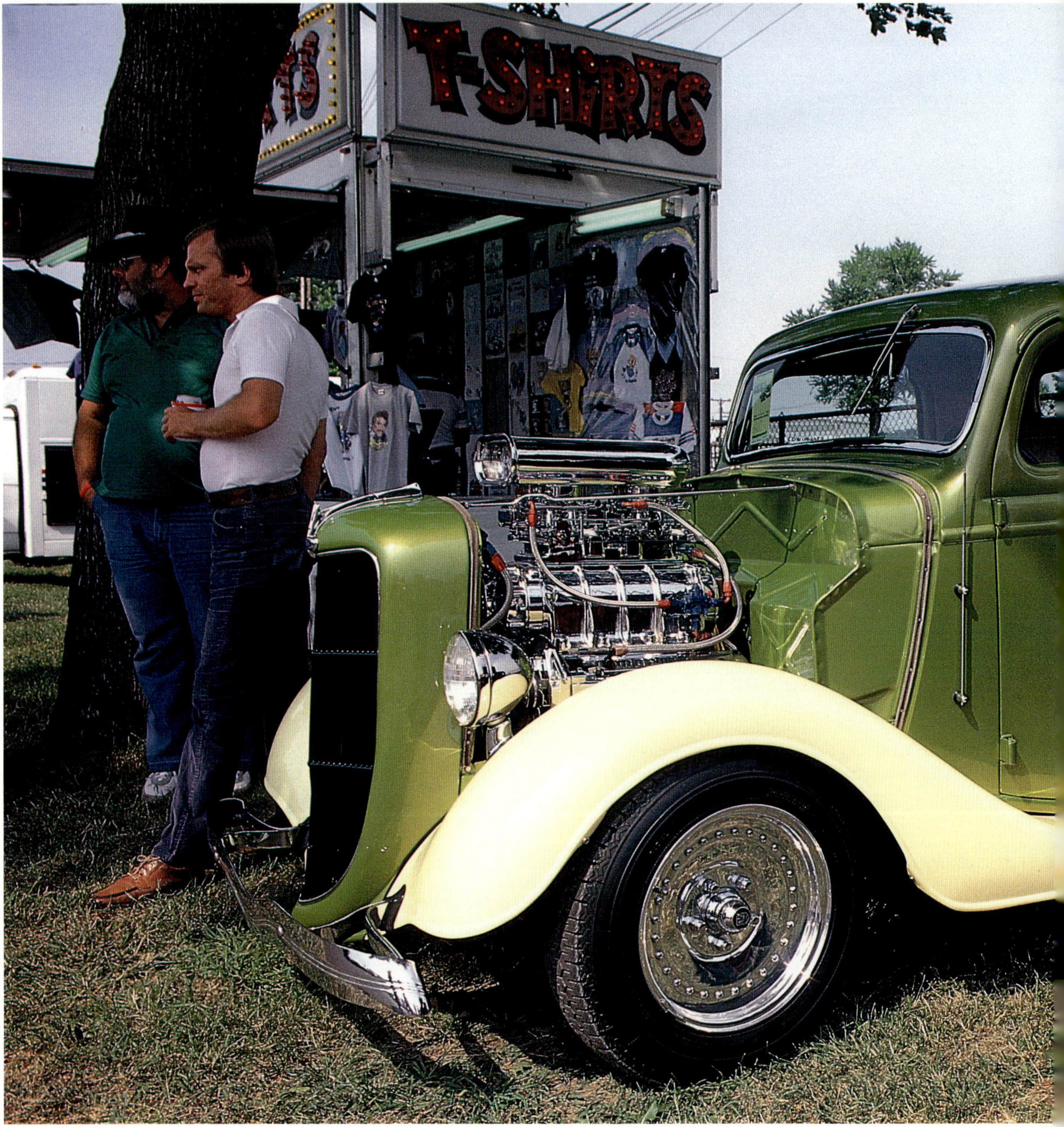
T-SHIRTS

Oldies But Goodies

From a lowly work machine has evolved a highly developed segment of the automotive enthusiast world: the street rod pick-up. The term 'street rod' came about in the late 1960s and early 1970s with the resurgence of interest in older, generally pre-war automobiles. It represented a conscious effort on the part of the enthusiasts involved to shrug off an image previously associated with the words 'hot rod'.

Back in the early Fifties hot rodders were generally considered to be the irresponsible youth group, performing dangerous stop light antics in stripped-down, hopped-up jalopies and racing on public roads after dark in defiance of the police. While no doubt based on a certain amount of truth, this image did nothing to help the more responsible, family-orientated hobby that developed many years later.

I'll huff and I'll puff and I'll blow you away! With a supercharged V8 for all the world to see, this two-tone 1936 Ford sure looks ready for action

A street rod was originally defined by the fledgling National street rod Association (NSRA) as a modified vehicle of 1948 vintage or earlier, which usually meant that a more modern drivetrain, suspension and brakes were fitted and the bodywork and interior were suitably rejuvenated. The choice of 1948 as a cut-off date has caused controversy ever since, especially amongst owners of 1949 cars and trucks who have found themselves confined to the spectators car park at major street rod events!

The appeal of a street rod pick-up is obvious. It is the combination of old and new; the stunning visual impact of a character vehicle that is so far removed from modern-day design theories yet so *slick*; a truck that is so obviously the result of the owner's handiwork and taste. But the amount of hard work and skill involved in rebuilding a 50-odd year old commercial vehicle to such exacting standards should never be underestimated.

There are several different styles of street rod, from the nostalgic Fifties look often applied to early Ford products, to the currently favoured 'Pro-street' approach featuring high-horsepower engines and ultra-wide rear wheels within the confines of the original bodywork. A huge parts industry has developed around the hobby and hundreds of street rod events take place every year in the USA.

What could be more fun than cruising in a street rod pick-up?

Left and overleaf
In the world of hot rodding, some cars become so well known they attain star status. The '32 Ford coupe in American Graffiti *spring to mind, likewise ZZ Top's Eliminator '33 Ford 3-window and* The California Kid, *a '34 Ford coupe driven by Martin Sheen in the 1970's TV movie of the same name. But when it comes to street rod pick-ups, Jim 'Jake' Jacobs' 1929 Ford Model 'A' is surely the all-time rolling legend. Jake teamed-up with Pete Chapouris many moons ago to form Pete & Jake's Hot Rod Repair, and in the late Seventies this marvellous old metallic truck served both as a rolling billboard for the business and as Jake's daily driver. The 'A''s original chassis was boxed and fitted-out with a Super Bell dropped tube front axle incorporating P & J's 4-bar links and Mustang disc brakes, along with a Halibrand quick-change rear end with coil-overs and ladder bars. Motivation was supplied by a 'mild' 327 cu. in. Chevy V8 mated to a Turbo 350 auto trans. The previous Metalflake paint job was tastefully re-sprayed, though the black tuck'n'roll style interior, as stitched in 1967, was retained. Sitting 'just right' on classic American 5-spoke wheels and radial tyres, the truck looks as impressive today as it did in St. Paul, Minnesota, way back in 1979*

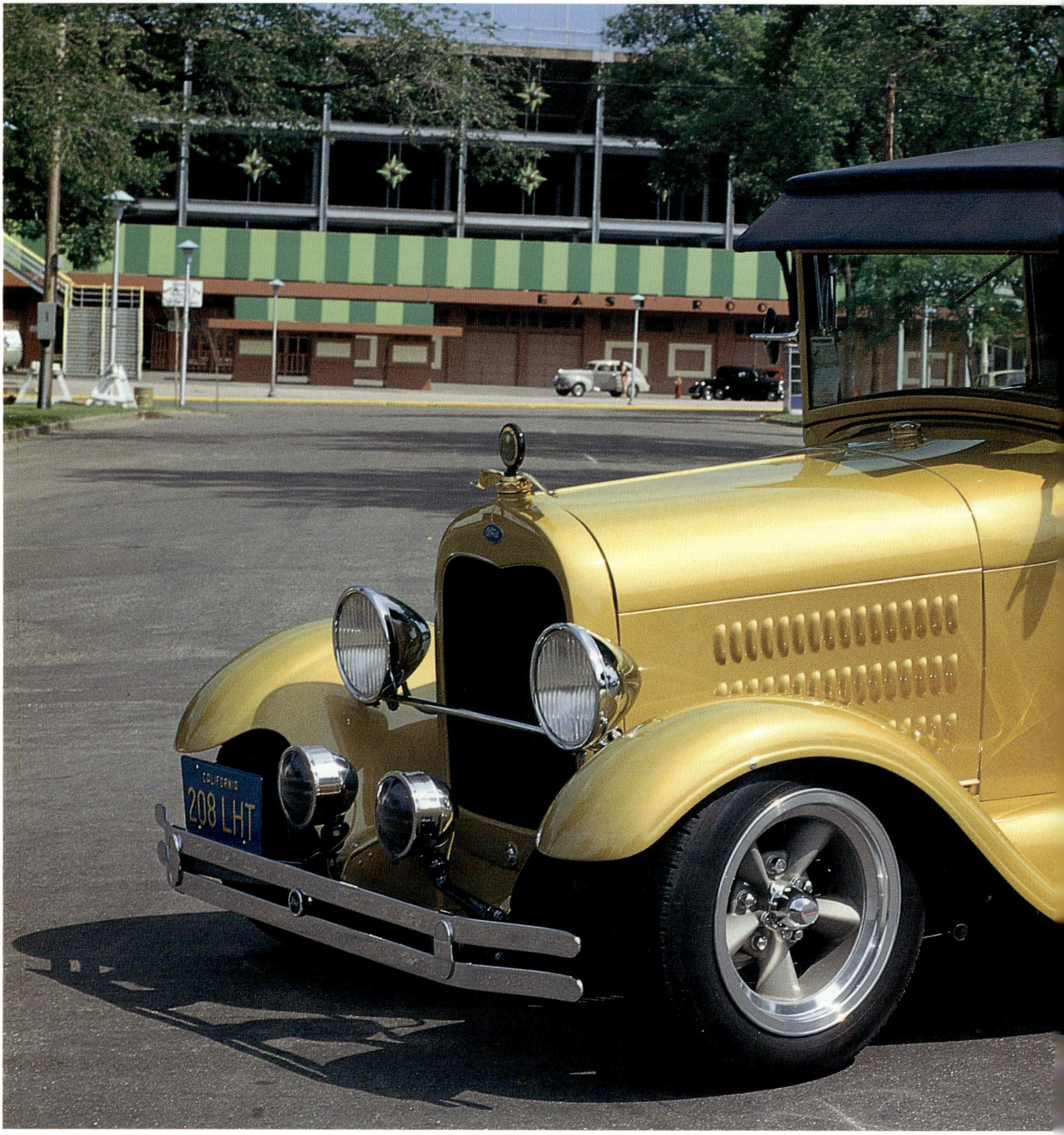
CALIFORNIA
208 LHT

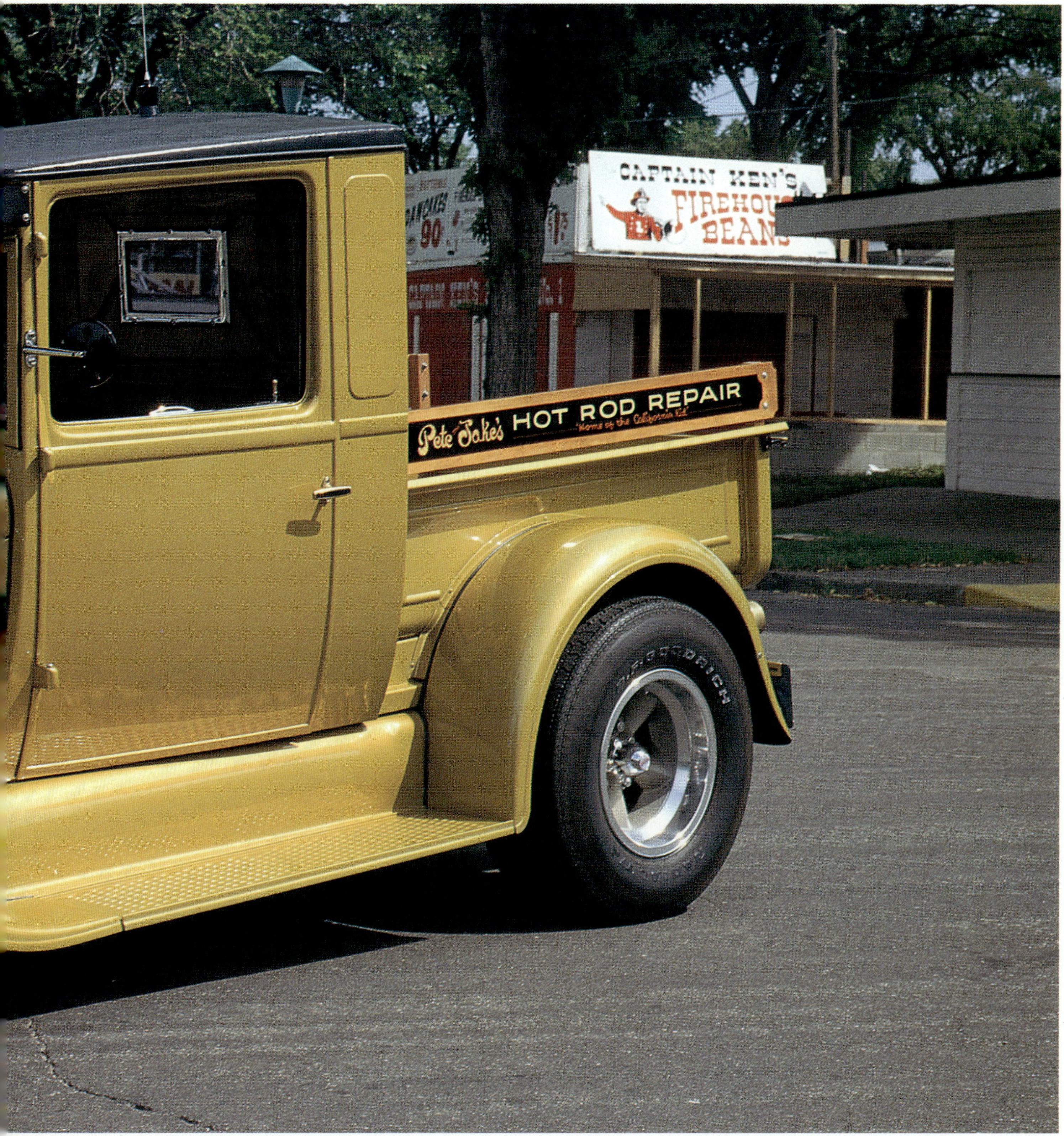
CAPTAIN KEN'S
HOT ROD REPAIR

Bright Model 'A' Ford stands out amongst the late-model clone cars of Oklahoma City. Body mods appear to be limited to a liberally louvred tailgate, but the truck's dumped-in-front stance, coil-over shock-suspended rear end and dual exhaust pipes leave the onlooker in no doubt that there has been a major modification under the vintage hood. Indeed, a lot of big boys would love this Tonka toy for Christmas...

Above and overleaf
One super-smooth 'hi-tech' street rod. Every panel on this 1932 Ford roadster pick-up has been relieved of all unnecessary hiccups in an attempt to 'out-smooth' the competition at the 1991 NSRA (National street rod Association) street rod Nationals, held in Oklahoma City. Modern-day rectangular headlights are mounted low and moulded into the characteristic grille shell, with its hand-crafted aluminium insert.

Similarly, the rear view mirrors are blended into the cowl, while the hood sides now incorporate reversed scoops in preference to the original vertical slats. The truck sits perfectly with its nose 'in-the weeds' thanks to careful setting-up of the independent front suspension and the very noticeable difference in front and rear tyre profiles. Wheels are the ever-so desirable Halibrands with triple-eared knock-offs. What a slick statement!

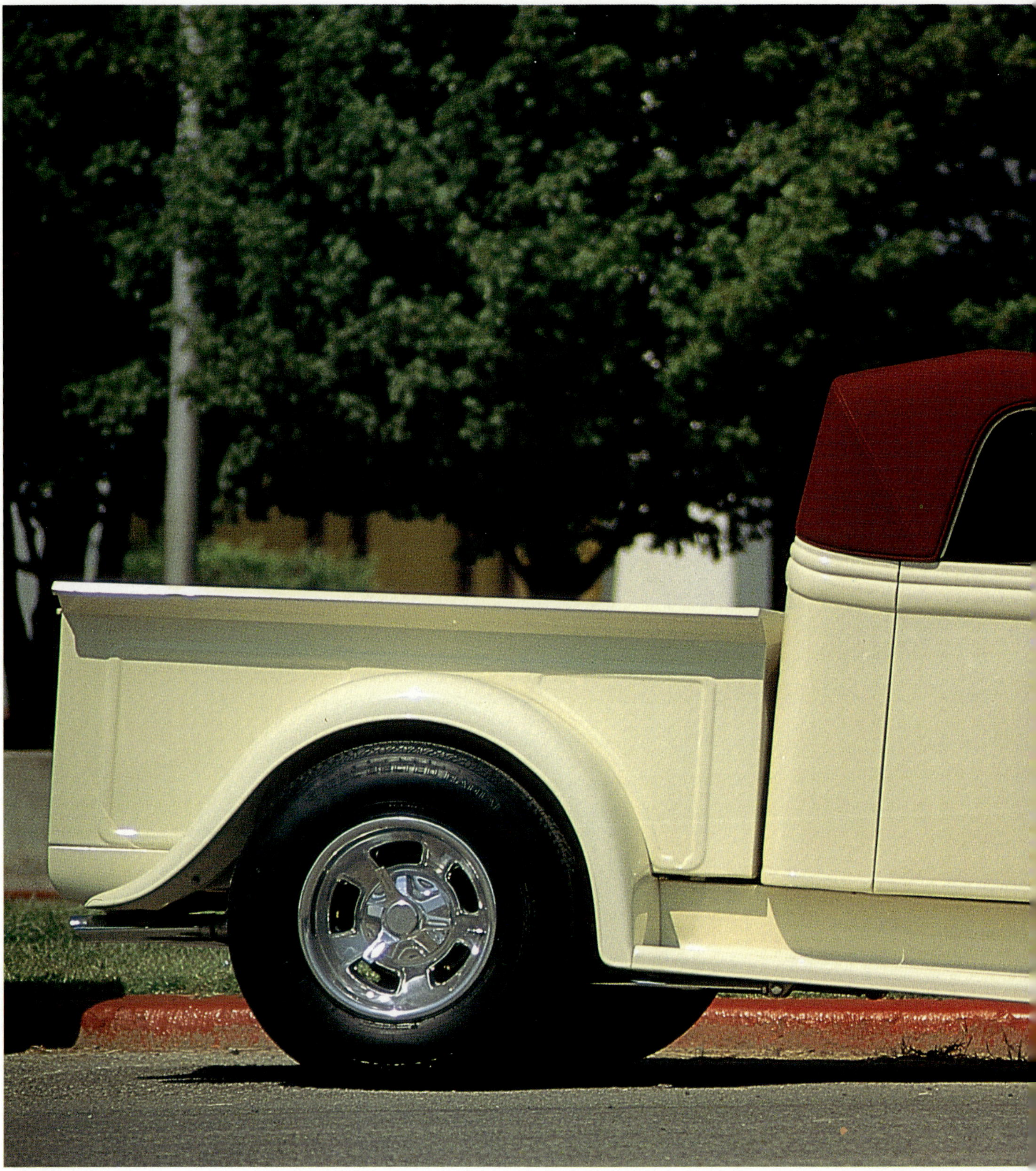

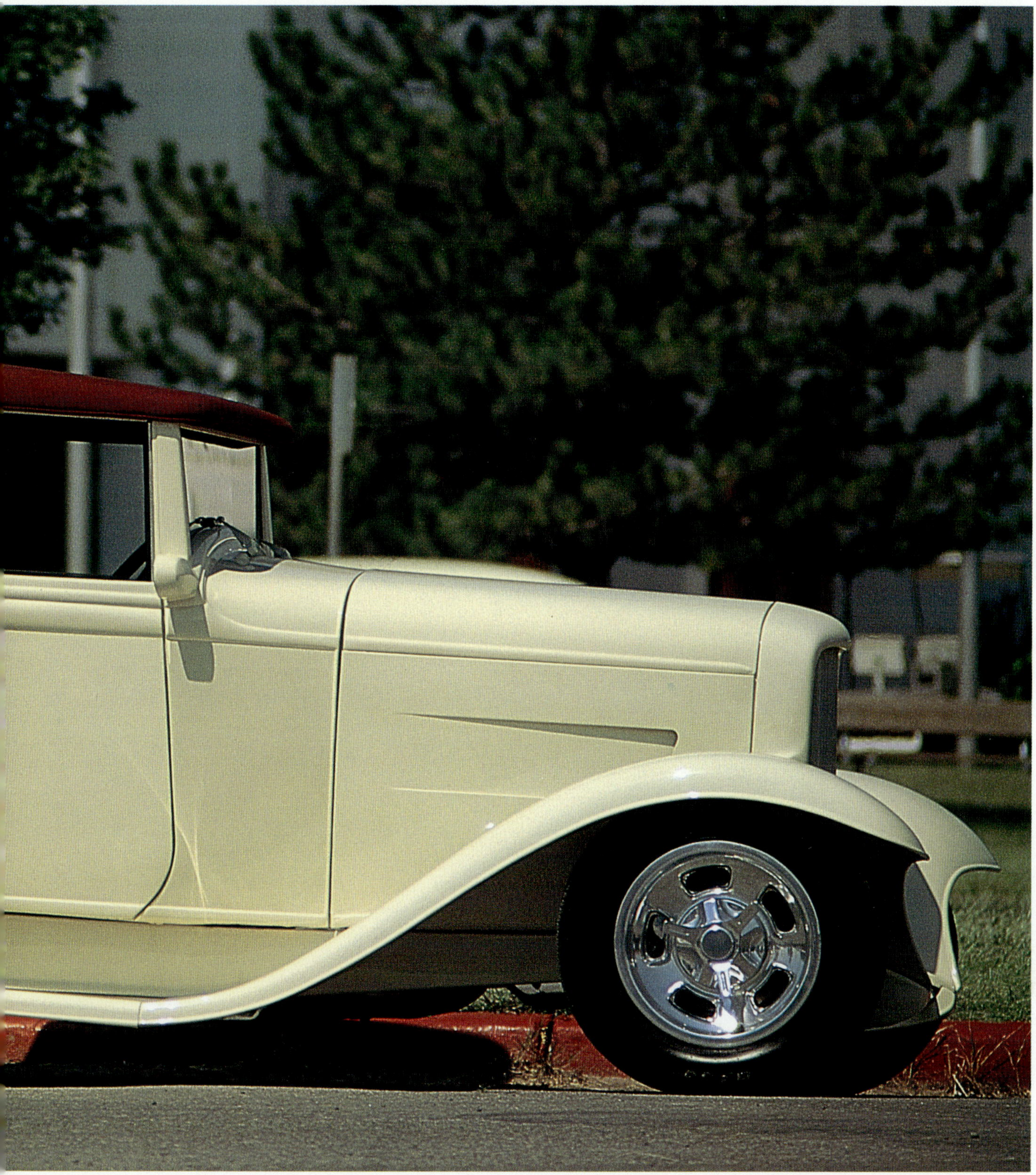

Time warp: Straight out of the Fifties, this '32 Ford closed-cab pick-up is pure nostalgia. Period perfect in every way, only the 'For Sale' sign in the windshield gives the game away by describing the modern-day Chevrolet V8 drivetrain. 'Wide whites' on red rims with chrome hub caps and beauty rings do much to create a traditional hot rod flavour in conjunction with the flat black primer paintwork, while the dropped headlight bar, the '34 Ford bumper and the louvred hood panels do absolutely nothing to spoil the effect. One can imagine this truck cruising Smalltown, USA around 1955, or making the long trek out to the dry lake beds at El Mirage with a belly tank racer in tow...

Long wheelbase 1934 Ford stakebed combines elements of both old and new, street rod and competition challenger. Massive aluminium wheeltubs cover huge rear tyres, while Pro-stock wheelie bars shows the driver, Eric Peratt, really means business. The new-age checkered paint scheme is more akin to the California mini-truck scene, though such ideas are gradually finding their way onto street rods all across the USA. This beast serves as a promo' vehicle for **Pinkee's Rod Shop**

Top-chopped and smoothed-out all over, this 'hot red' '37 Chevy is the quintessential Nineties street rod truck, complete with 'Pro-street' wheels and tyres and a big-block motor. Check the side-opening hood, the absence of all trim, door handles and bumpers, the frenched aerials in the cowl, the colour-coded milled aluminium mirror – yeeow!

7-up in the bed of a lipstick-pink '37 Ford, a home-built roadster from Nebraska. Indeed, what better way to see some 8000 street rods at the NSRA Nats?

A real home from home, this old Chevy certainly has rustic charm. Curiously, shack and outhouse custom campers were quite popular in the Seventies, though it's doubtful whether there were any more authentic than this DIY job with its roof shingles and shuttered windows. (It should be noted that the owner was checking into a Motel 6 as this picture was being taken)

Another creative camper, this time in aluminium. Though lacking the old-world charm of the one opposite, full practicality is ensured with a 'diamond-studded' rear door and a roof rack to carry all those essential items for the great outdoors

Above
Wooden ya know, you just can't keep a good craftsman down. Product loyalty may be big in the States, but this 'Hillbilly' kind of treatment all but died-out during the last decade.

Right
Over half a century separates these two speeding trucks; one high and mighty, the other hunched low to the ground in the traditional hot rod rake

25

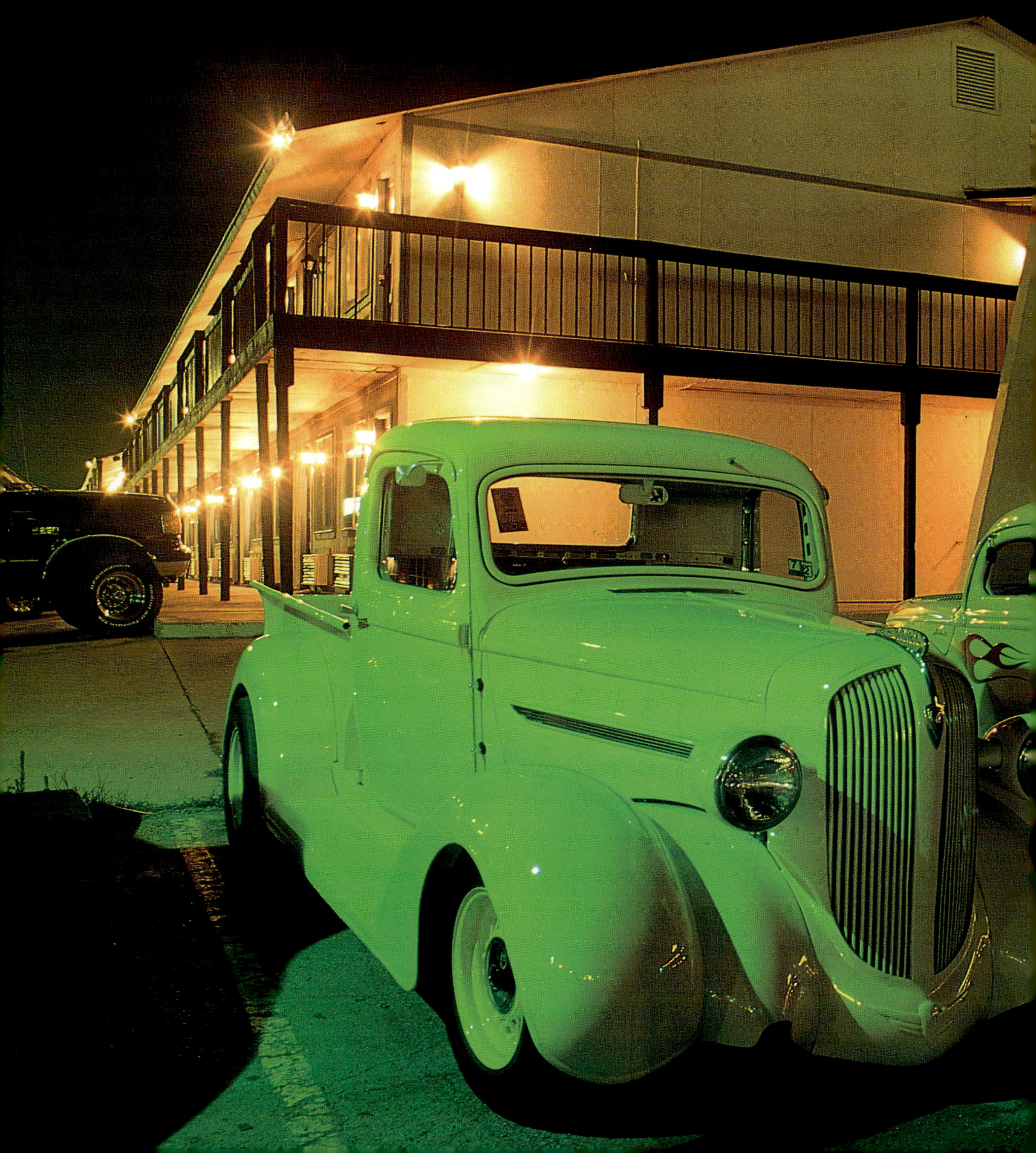

Pictured under the eerie lights of an Oklahoma City motel, these two elderly pick-ups, a '38 Plymouth (left) and a '40 Ford, cool-off after a sweltering mid-summer's day

Refreshingly different in aqua and white, this '35 Ford half-ton is an excellent example of modern re-styling. Although the cab has been subtly chopped and the hoodsides smoothed over, the rest of the bodywork is pretty much as Henry Ford intended. Wonder what he'd make of that new-wave graphic paint?

Quick, grab a fire extinguisher! Flames will never go out of fashion, and this chopped 1937 Plymouth will remain in vogue for several years to come

29

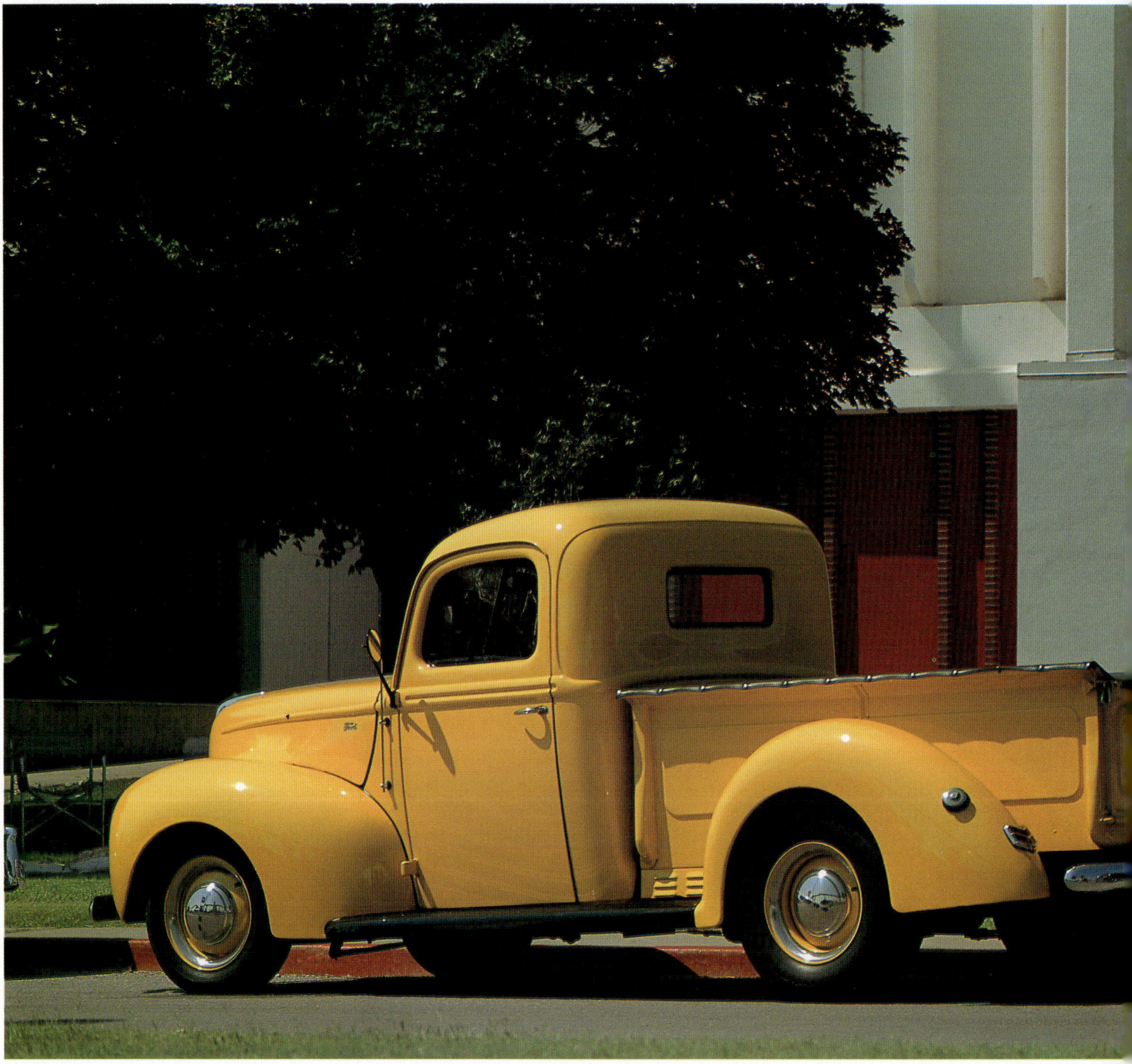

Left and above
Styled after the passenger car line, the 1940 Ford truck is an undisputed classic which can benefit from the restrained, 'resto-rod' treatment. Only a lowered front end, slightly fatter rear wheels and tyres and a fresh lick of paint separate this stylish '40 from a stocker – though you can bet its owner could point-out several other changes beneath the skin

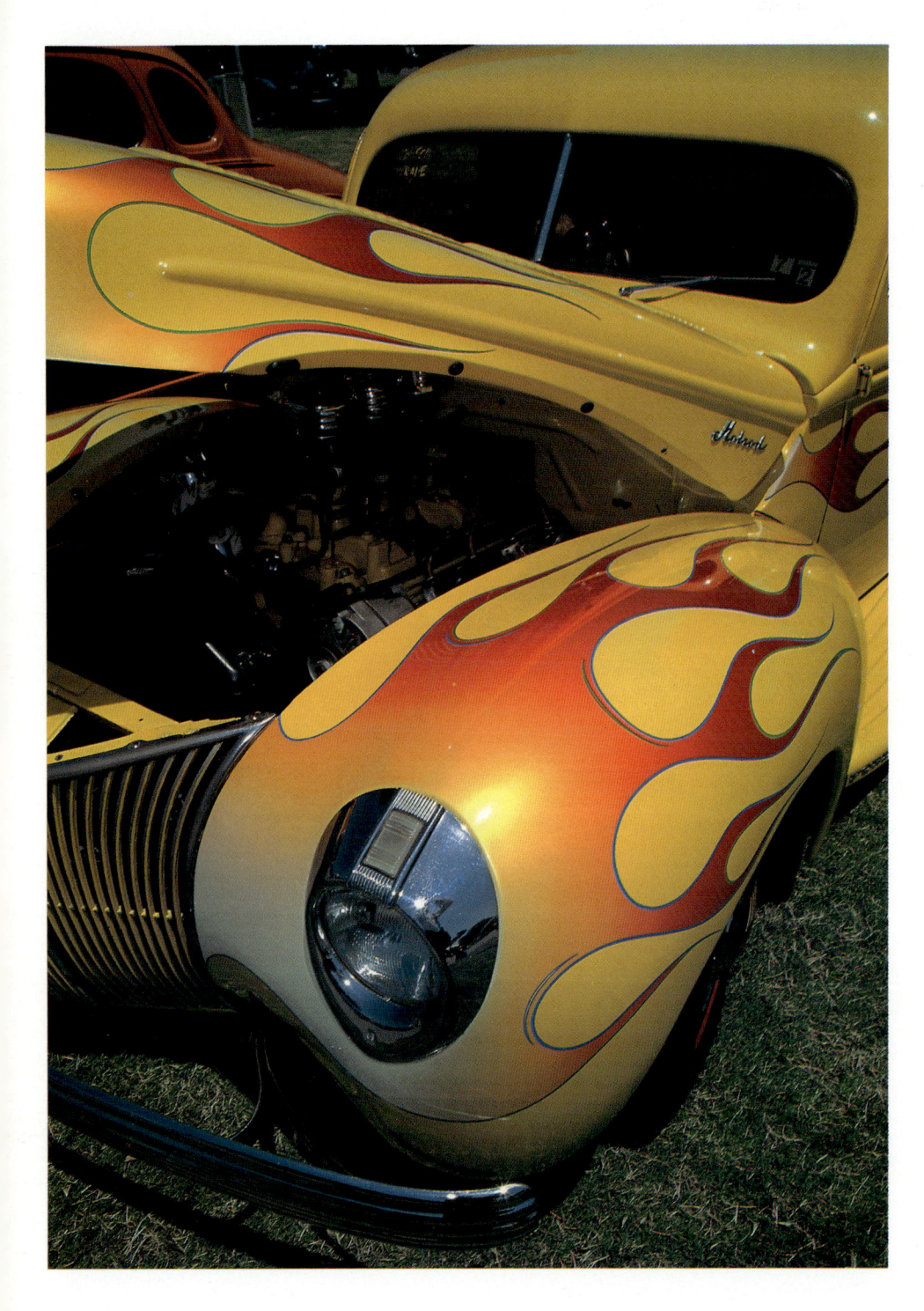

The wonderful thing about Henry's 'Forty', truck or car, is that whatever traditional modifications are made to the model, it always looks 'right-on'. Three carbs on a Chevy V8 lurk deep inside the engine bay on this Candy flamed, chopped example

An altogether different approach, this time on a '41 Ford. The owner, Harry Shepherd from Dallas, Texas, has taken his inspiration from the ever-growing mini-truck scene and produced a really eye-catching phantom roadster pick-up. Ford produced their last topless truck in 1932, but a deft slice by Harry ensures the style lives on. The decapitated cab has been given over-size rear fenders, slimline running boards and trick lavender scallops over pearl white paintwork. Modular spoked wheels echo the contemporary look, but the original chromework ensures this fun truck never loses its true identity

It is said they like 'em bigger in Texas. Danny Greenhaw from Kaufman, Texas certainly does. The only way to own another truck like his bright red one will be to build it! This 1940 Ford crew-cab is what's known as a 'phantom' body style, or one that was never produced by the manufacturer. What's more, the work that will have gone into producing this unique 'Supercab' should not be underestimated. The rear section of the original cab has been cut off and another section (incorporating a cut'n 'shut window opening) expertly grafted in, thereby producing the crew-cab effect. The chassis and wheelbase was also extended in order to accomodate the original stepside bed. Unveiled at the Nationals in '91, Danny's flawless '40 will no doubt serve as inspiration for many

Attention to detail is what makes or breaks any customised vehicle. The owner of this '39 Chevy light truck has chromed, polished or painted every single part on his chosen small-block Chevy engine.

No book about custom pick-ups would be complete without at least one Willys. This pint-sized 'Pro-Suede' truck probably spent its early years as a drag racer. Many refugees from the 'Gasser wars' of the Sixties are now enjoying rejuvenation as street rods

Beautifully-finished 1946–48 model Studebaker makes for an unusual street rod truck. And you can bet that Candy paint has earned more than a few trophies for its owner

Above
Mr and Mrs Paskiet in their 1941–47 Dodge WC Series half-ton. Exact identification is difficult from this candid shot as Dodge produced this model from '39–47, but the cowl-mounted parking lights first appeared in '41

Right
A variation on an old joke. Usually the wife threatens to leave if the husband refuses to sell the truck

Left
How about this for an unusual set of louvres? Hot rodders are continually striving for the kind of original idea that this Chrysler product displays

Below
Cartoons and murals were once very popular but are rarely seen now. Who knows what the significance is here? Perhaps the owner is a pig farmer...

Above and right
This is one of several 1946–48 Fords that utilize passenger car front end sheetmetal in conjunction with a half-ton stepside bed from another source. Moreover, with the complete removal of the top, a 427 cu. in. Chevy motor and state-of-the-art features throughout, this former '47 four-door sedan must be the ultimate phantom roadster pick-up. The tilt-bed truck was built by Roy and Paul Poindexter of Oklahoma City, though, the original inspiration is said to have come from a plastic scale model built by Californian modeller, Dave Hill

Following spread
Chopped, lowered and colour co-ordinated for an eye-popping reaction. Like it or not, this '48–50 Ford F.1 represents a truly creative package that is highly indicative of the 'dare to be different' approach currently propounded by Hot Rod and other enthusiast magazines

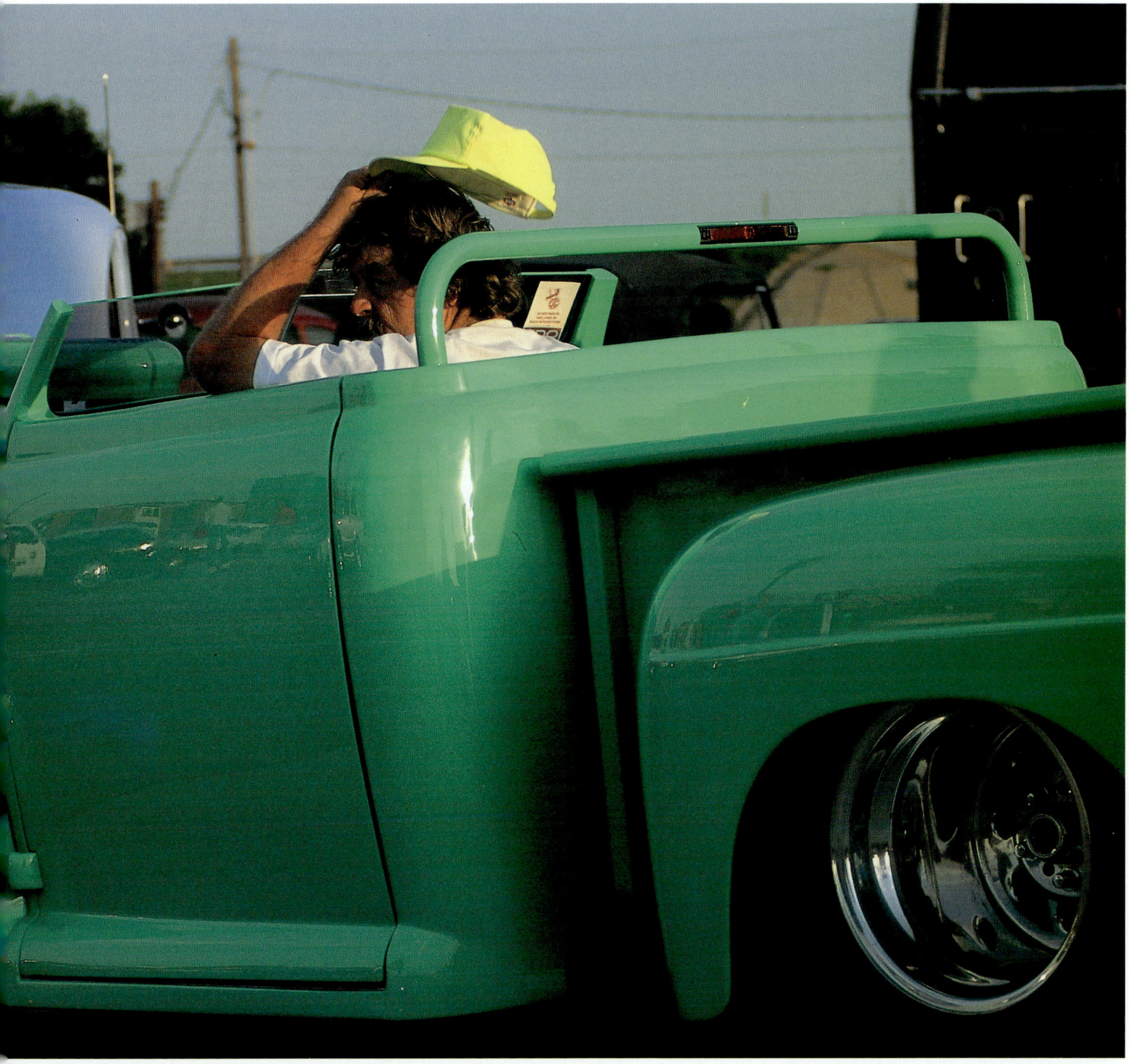

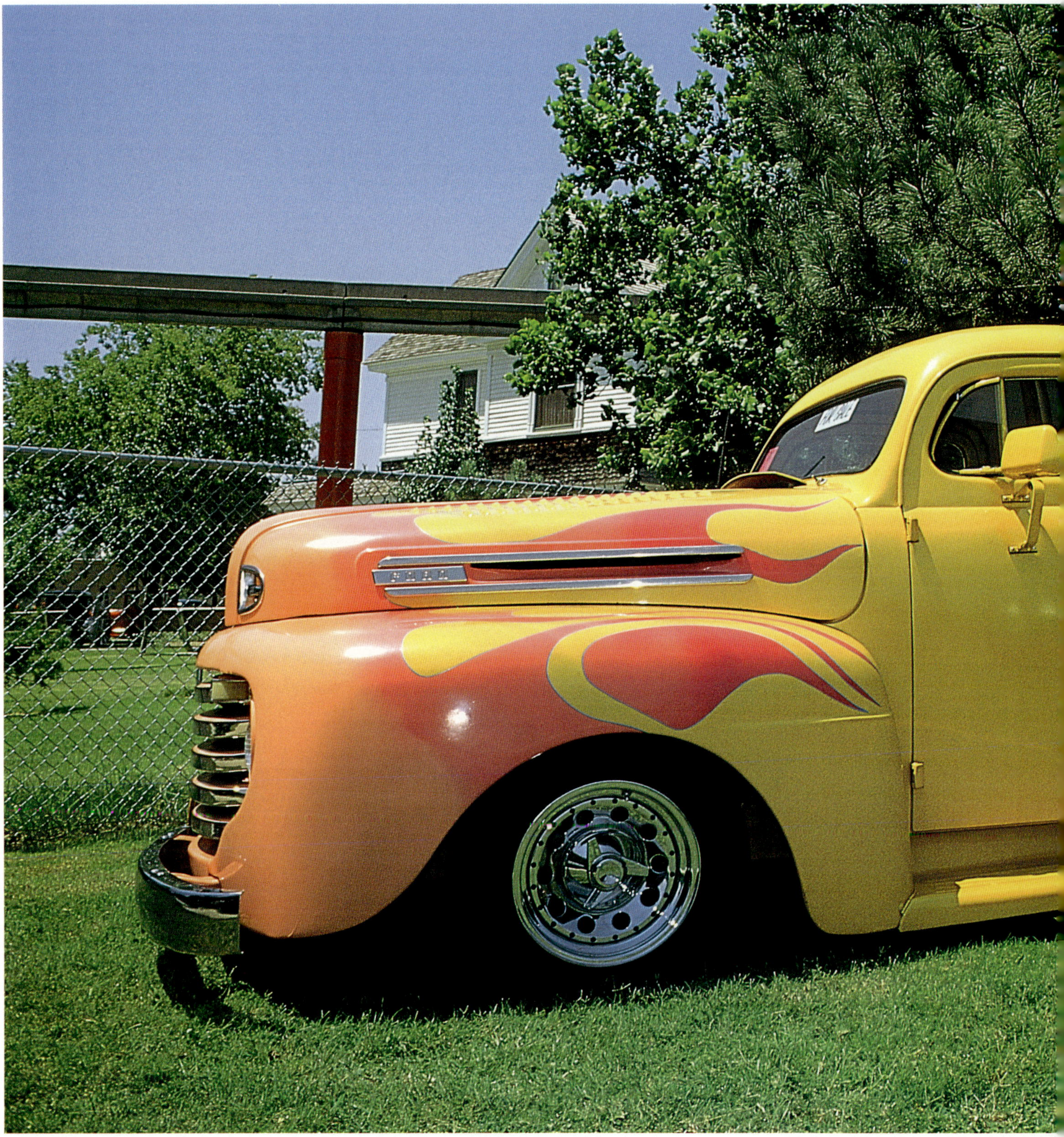

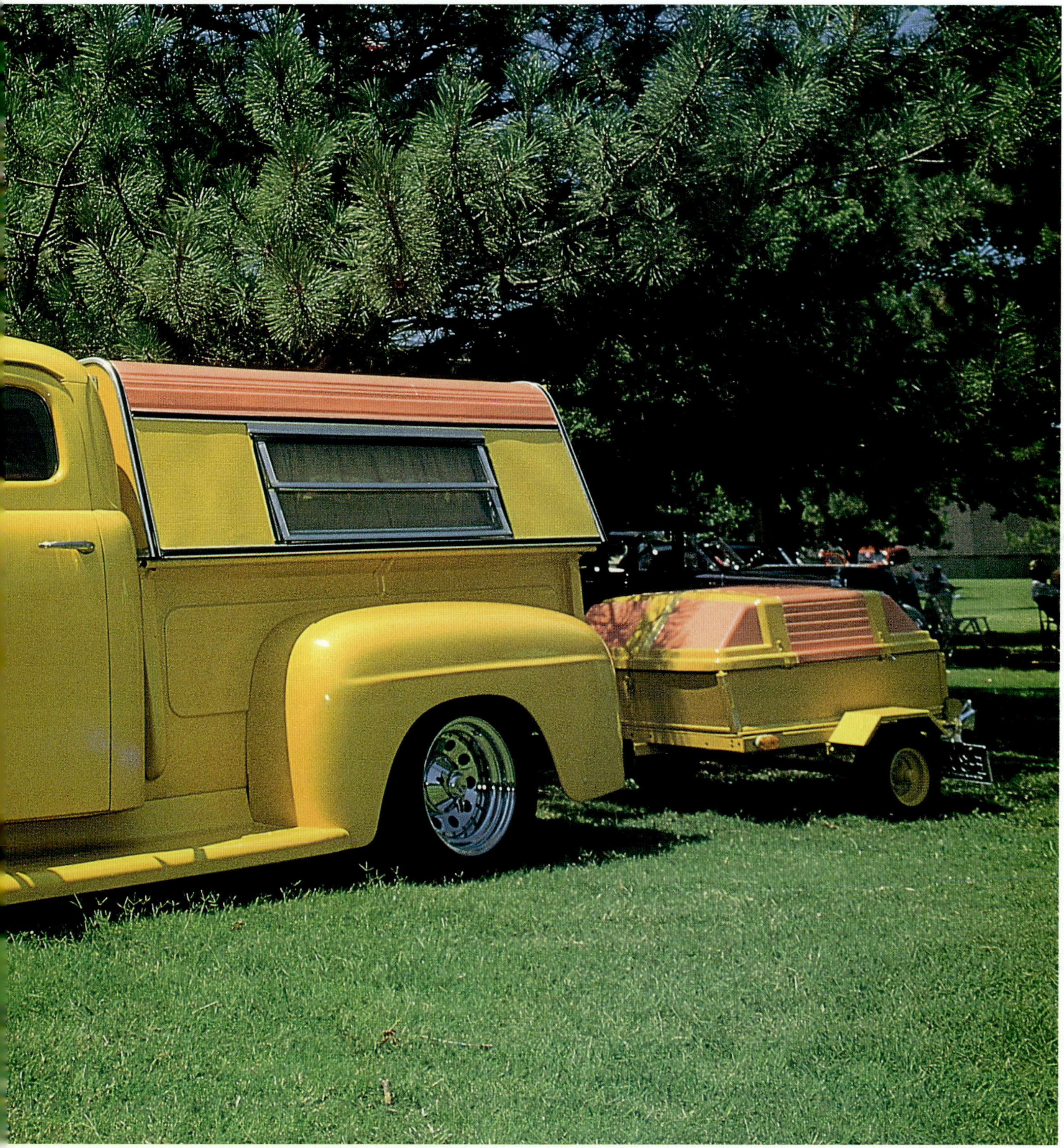

44

Left
Radically chopped F.1 clearly indicates how tastes have changed. Once considered the ugly ducklings of the pick-up world, a shift in attitude during the Eighties brought widespread acceptability to even these forgotten Fords. The owner of this tangerine dream obviously thought carefully about the overall effect before starting the project

Above
Simple but effective flames on a '46 Ford commercial (licenced 46 HEAT)

45

Above
Flat-bed trucks come in handy for a wide variety of tasks, but it's not often you see such a well-detailed version. Featuring a checker plate bed attractively edged with polished wood, this late Forties Chevy rolls on widened Camaro Rallye wheels

Right
A face only a mother, or a street rodder, could love. Flamed Forties Chevy is brimming with unusual features, most notably the dark Perspex-covered flush headlights which lend a feeling of menace. Shot at the KKOA (Kustom Kemps of America) Nationals in Missouri, this vehicle effectively makes the transition from street rod to kustom

Above
One super-smooth customer. Shapely front end has been relieved of all chrome trim and the original fender-mounted headlights are gone, replaced by a pair of driving lights sited inside the grille. A tastefully chopped top and a frameless windshield complete the transformation from Forties farm hack to Nineties-style rod

Left
Original '41–47 Chevy front end. Factory 'bullet' headlights retain piggy-back parking lights, and see how the windshield frame hinges forward for ventilation. The proximity of the chrome bumper to Mother Earth, however, suggests that this is a subtle resto-rod rather than a straight restoration

'Lowered-to-the-max' is the only way to describe this '47 Ford. It was driven from Arkansas to OK City (with its original licence plate in situ) to be part of the NSRA's annual happening

The miniscule Morris Minor is a hit on both sides of the sink. This one has been expertly converted to a stepside and, what's more, you can be sure there are a lot more than 1000 cc's under the bonnet...

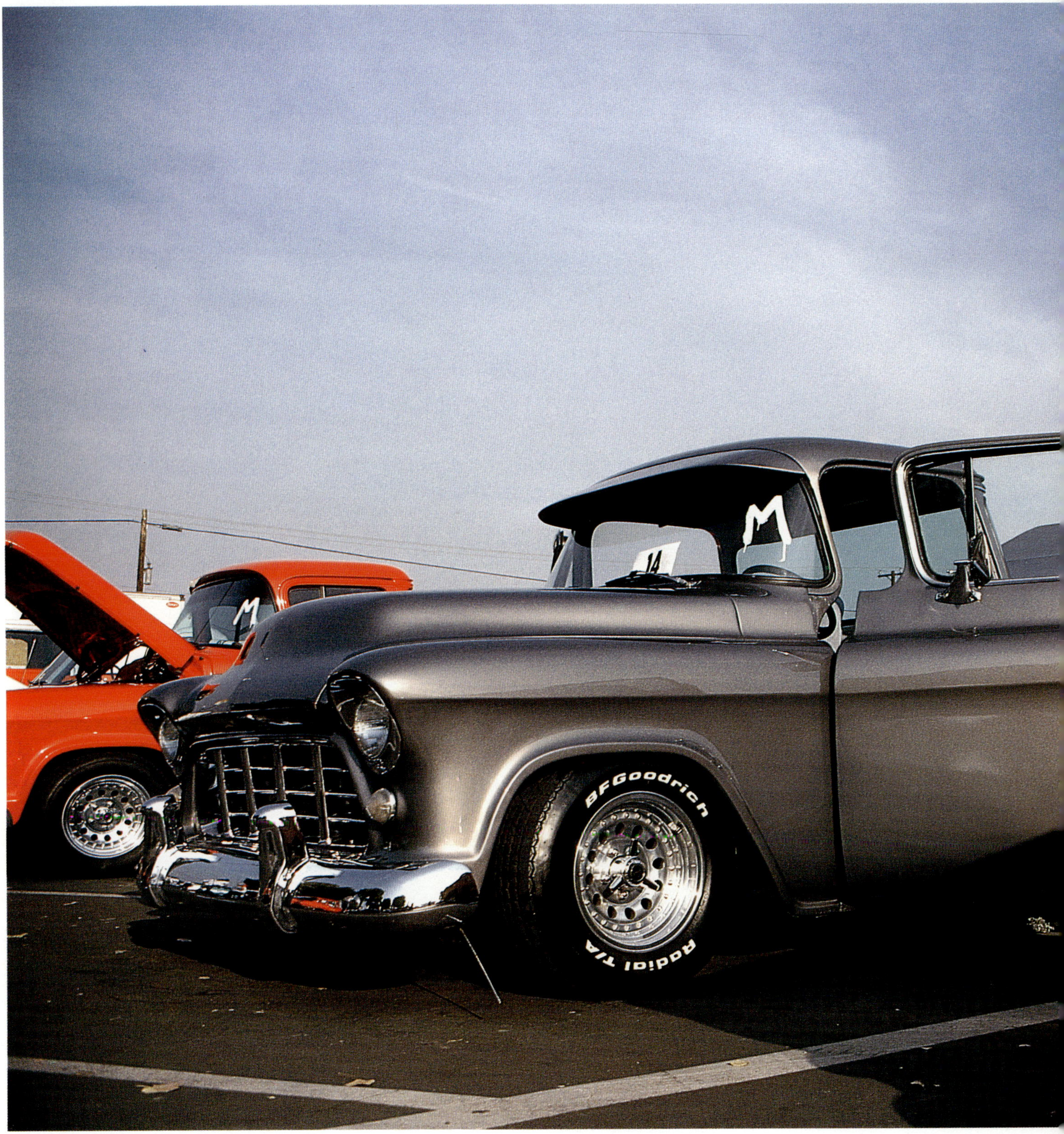
14
BFGoodrich
Radial T/A

Fabulous Fifties Rigs

Following the cessation of war-time hostilities, automotive designers pushed ahead making the most of new manufacturing techniques and materials developed during the war years. Although not so style-conscious at first, commercial vehicles soon followed Detroit's passenger cars in displaying a smoother, more integrated overall design.

The 'separate fender' look began to disappear in the early Forties in favour of a fatter, faired-in, more streamlined style, and this trend continued through the Fifties. Truck designers became more aggressive in their attempts to woo potential buyers making

Sharp '56 Chevy stepside with smart metallic grey paint, period sun visor and lots of chrome. The truck has been extensively up-dated by substituting a late-model Camaro engine and complete front subframe. This has the advantage of independant suspension with motor mounts, power disc brakes and steering already correctly set-up from the factory. A smooth ride, likewise performance and reliability, is guaranteed

stylistic changes every year.

'Wraparound' windows were introduced in trucks by the middle of the decade, likewise gaudy ornamentation, following the lead set by the passenger car market. Pick-ups became stylish as well as practical in an era dominated by fins and chrome and heavy-breathing V8 engines.

But that is not to say these trucks were less durable than their predecessors. Far from it. Even now, over thirty years later, large numbers of Fifties American pick-ups are still performing valuable daily service. They have not yet achieved the collectors value of many pre-war trucks, however, and that is one of the main reasons for their continued appeal amongst customisers.

With running gear that is still reasonably practical for use on today's roads and engines, often optional V8s, which are still acceptable, Fifties trucks are both affordable and available. They respond well to mild customising and will accept late-model high-performance engines and transmissions with relative ease thanks to their cavernous engine bays.

Certain marques like the Ford F.100 and Chevrolet half-tons have become so popular that a reproduction parts industry has developed around them whereby it is possible to obtain everthing from a purpose-built independent front suspension system to the most obscure original part through a phone call to any one of many specialist companies.

Long live those fabulous Fifties pick-ups!

Left
The born-again Rod and Custom Magazine *'Dream Truck', probably the most famous custom pick-up in the world. Featured in over 90 magazines over the years, Spence Murray's radically kustomised 1950 Chevy was purchased back in '53 as a daily runaround-cum-project vehicle. Spence, the editor of the fledgling R&C at that time, substituted a new '54 cab for the original and had it chopped, channeled and sectioned for a series of articles, arranging for several of the best-known 'shops of the period to perform ever more radical modifications. Valley Custom, Gene Winfield and the 'King of the Kustomisers', George Barris, all played a part in creating this unique design which also featured the first recorded Chevy V8 engine swap. Then, in October 1958, tragedy struck. The truck was virtually destroyed while being towed to a show. The wreck passed from hand to hand over many years before the eventual restoration by master bodyman, Carl Green. Now owned by Kurt McCormick, the 'Dream Truck' once again cruises the boulevards in all its 1950's glory*

Below
Shades of the Sixties: Somewhat psychedelic '48–53 Chev' serves as 'art-chaser' for an establishment in Albuquerque, New Mexico

Above
Who could possibly fault this low-riding Chevy half-tonner? American Racing 5-spoke wheels are a timeless addition to virtually any modified street machine, while the monochromatic metalwork, with frenched-in headlights and concealed solenoid-activated door locks, is as smooth as can be. It's all the reference you need...

Right
The face of Chevrolet's light truck of the early Fifties is generally thought to be far more attractive than its Ford counterpart

'Green Giant' is the apt name given to Dave Moore's 3100 Chevy truck. It first saw the light of day in '53 but looks and drives a whole lot better after Dave spent 10 loving years customising it. Extended wheelarches, chromed running boards, Cragar SS wheels and a fully trimmed interior featuring '68 Camaro seats are a few of the changes, while the bull-nosed bonnet hides a hopped-up 327 cu. in. small block Chevy V8

55

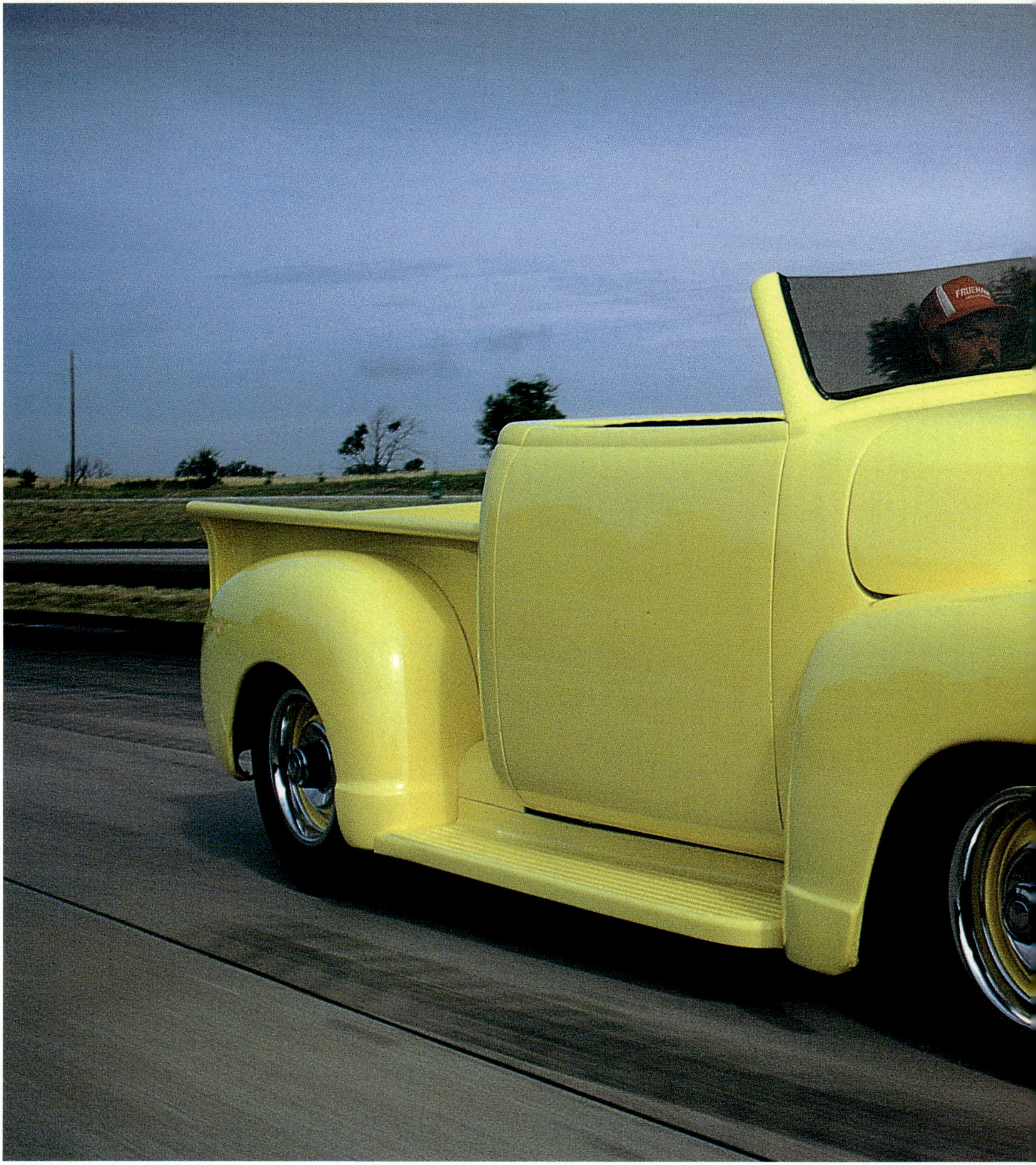

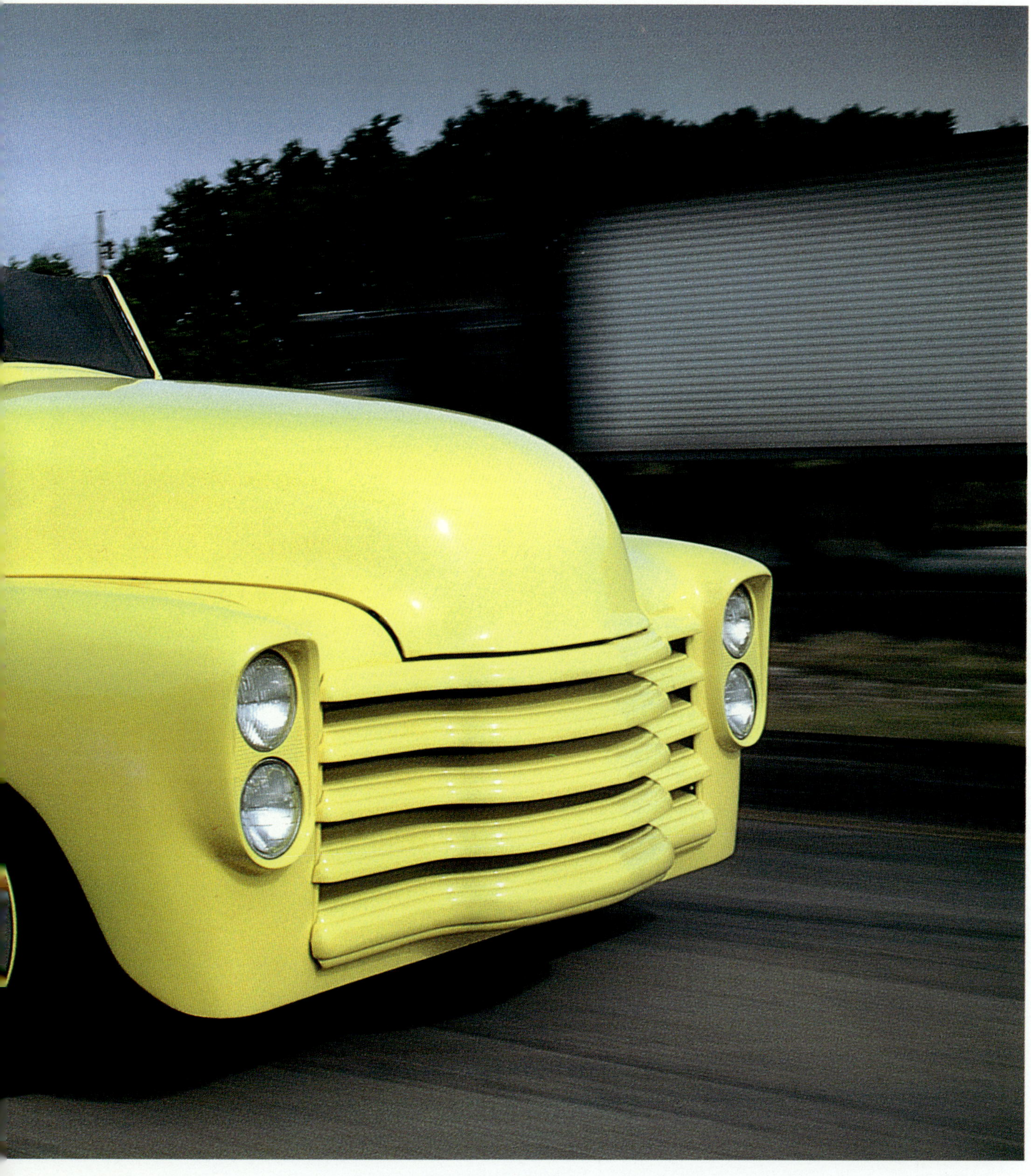

Previous page and above
Now here's a neat idea. Take one elderly pick-up, lop the top off, rid the body of all superfluous trim and paint it a nice bright colour and you could end up with a truck that's as trick as Fred Sullivan's low-down '51 Chevy. This Arkansas smoothie was built for low bucks with more than a hint of hi-tech custom style. Note the frenched-in quad headlights, the colour-matched Chevy Rallye wheels, the subtle forward rake...

The tailgate has been welded permanently shut and panelled bare. All chrome trim, including the unnecessary door handles, has either been removed or painted. The rear lights now reside behind a flush panel where the bumper would have been. Bang up-to-date with its semi-framed tinted windshield, Fred's fine ride proves you don't have to spend a fortune to get noticed

Above
A side-fitting spare wheel mount with indented fender is a desirable option on mid-Fifties Chevys. This one is enhanced by a colour-matched metal cover pirated from the back of a late-model custom van

Right
Classic Chevys congregate for an annual meet at Six Flags Magic Mountain, just outside Los Angeles. Amongst the great selection of vehicles on display in November 1989 was this '57 'big back window' Chevy 3100 (compare it with those further along). The 'cowboy's Cadillac' has been suitably customised with wire wheels, lowered front end and a powder blue paint job

61

Two in front, two out back and all four having fun in the sun. Wraparound glass provides a panoramic view of the road conditions for the driver of this 80 in. tall '56 Chevy

Words cannot describe the finish on this bed, suffice to say this is one truck that will never ever carry horse feed or lumber again

EL
MAR
Corona Extra
HUEVOS CON CHORIZO
LUNCH SPECIALS
TACOS AL PASTOR
SODAS
CERVEZA MEXICANA
WELCOME
PASE UD
3100 Chevrolet

Mildly customised '57 Chevy, pictured in Buena Park, SoCal, is typical of many in everyday use across the USA. Still fairly plentiful and reasonably cheap to buy in the mid-west states, the biggest problem, as with restoring any commercial vehicle, is removing all the dings and dents accumulated throughout a long, working life. Not that there's much evidence here. With a set of chrome rims and a freshening-up of the many pieces of trim that characterize the model, this time-piece turns heads all over town

Contemporary aftermarket wheels seem a little out of place on a classic Fifties pick-up, but then, as they say, beauty is in the eye of the beholder

Right
Chevy cab has been the subject of much tender loving care. Although, in essence, it remains as factory supplied, the steering column has been replaced with a tilt version from a late-model car topped with a sporty aftermarket wheel. This combines the transmission shifter and turn signal stalk. Re-upholstered, re-carpeted and re-painted throughout, the appeal is obvious

Overleaf
In the weeds! Pictured in a wrecking yard on the legendary Route 66 at Tucumcari, New Mexico, this late Fifties gem looks remarkably complete considering the location. Judging from the optional jet plane hood ornament and trim, this is an up-market variant of the Model 3A half-ton Chevrolet. First introduced in 1958 and carried over to the following year almost unchanged, the 'Apache' (also a 'Parisian street ruffian') nomenclature was applied to fancier models available with either two or four-wheel drive and a choice of six cylinders or V8 engine

67

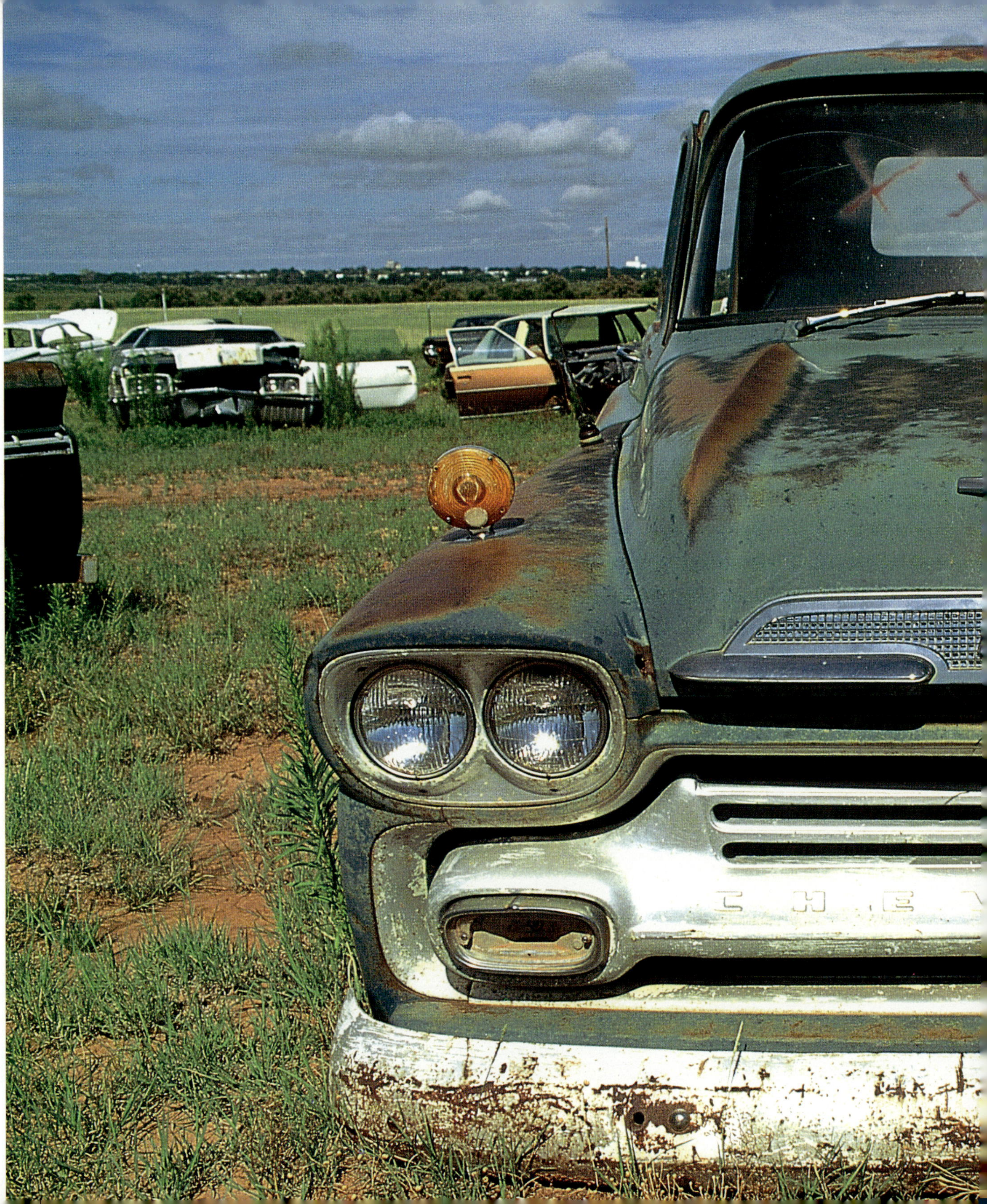

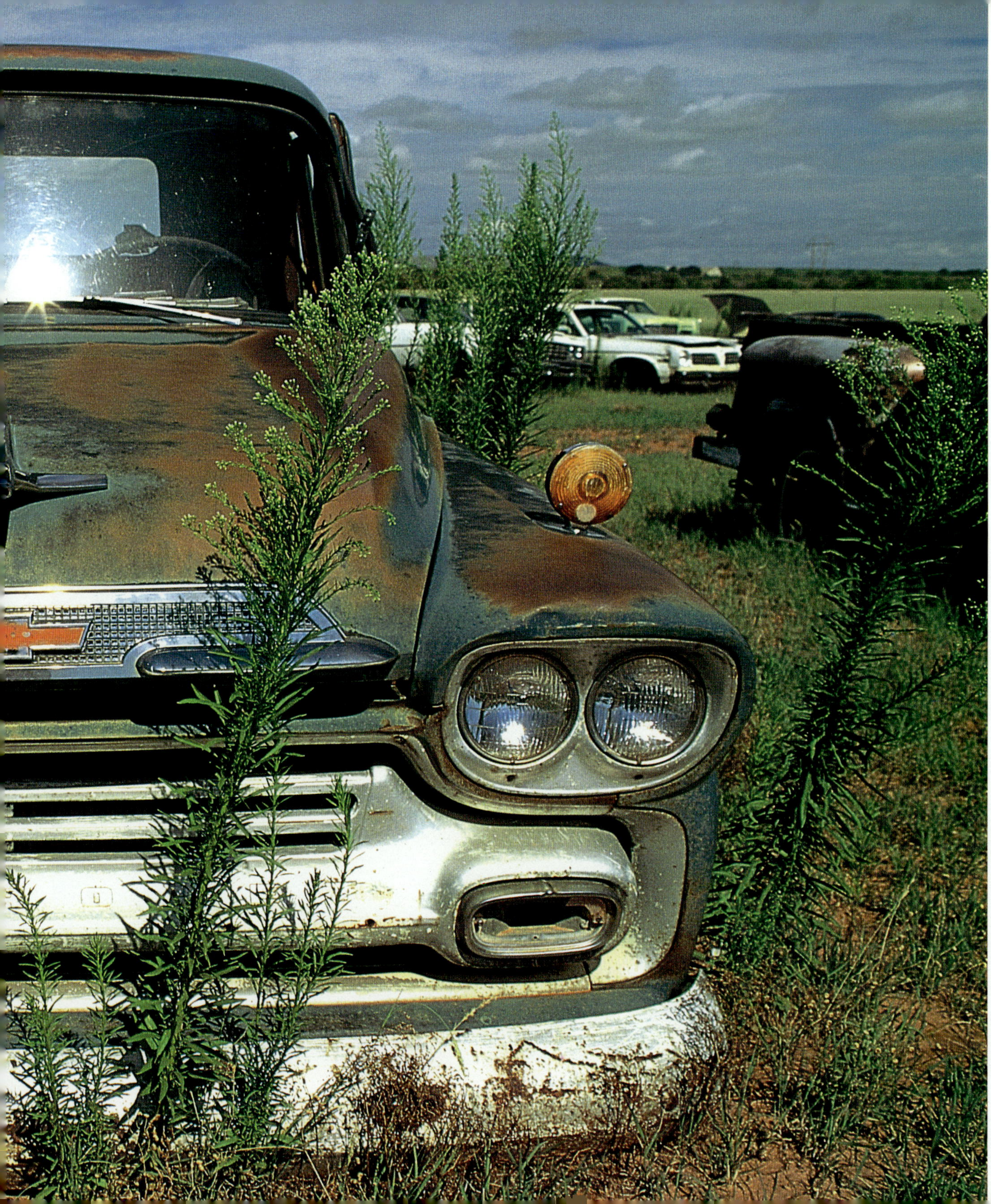

The fender flash denotes this is also an 'Apache' – but what a difference to the previous one. A great deal of time and money has no doubt been spent transforming this stepside into the stunner it is today. Chevrolet pick-ups of this period are unusual enough without needing extensive modifications to attract attention – on the surface, that is. Under the hood is a full-on 427 cu. in. 'Rat' motor

Stock-as-a-rock 'Jimmy' (GMC) truck epitomizes the extravagence of 1950's American auto design. GMC trucks shared the same basic tooling as their Chevrolet cousins. However, as this 1955 GMC 100 demonstrates, the egg-crate grille of the more common Chevy was replaced with the 1954 Oldsmobile-inspired monstrocity you see here. The cost of re-chroming this amount of brightwork must be horrendous but the effect is worth every cent when the overall restoration is as perfect as this one

Above
Whilst Chevy pick-ups from the Fifties are very popular, the sheer enthusiasm for the 1953–56 Ford F.100 is overwhelming. There are F.100-only clubs and numerous events organised throughout the US specifically for this model. A massive range of reproduction parts is available to ensure surviving examples remain in regular use. This flamed '56, lined-up with others in St. Paul, Minnesota, typifies the treatment so many receive

Right
Chopped, channeled and sectioned, this mid-Fifties Ford is truly 'rad-to-the-bone'! Whoever executed the complex body mods must have been some handyman, since there isn't a line out of shape nor a hint of filler

Left
Model 'T' Ford brass radiator and headlights is certainly unique on a Ford F.100. Maybe the owner really wanted a street rod?

Overleaf
Pretty pink paint belies the business-like powertrain beneath this F.100, captured during the night-time activities at the 1991 street rod Nationals. Unfortunately, the occupants were enjoying themselves far too much to talk 'spec', suffice to say it made all the right noises...

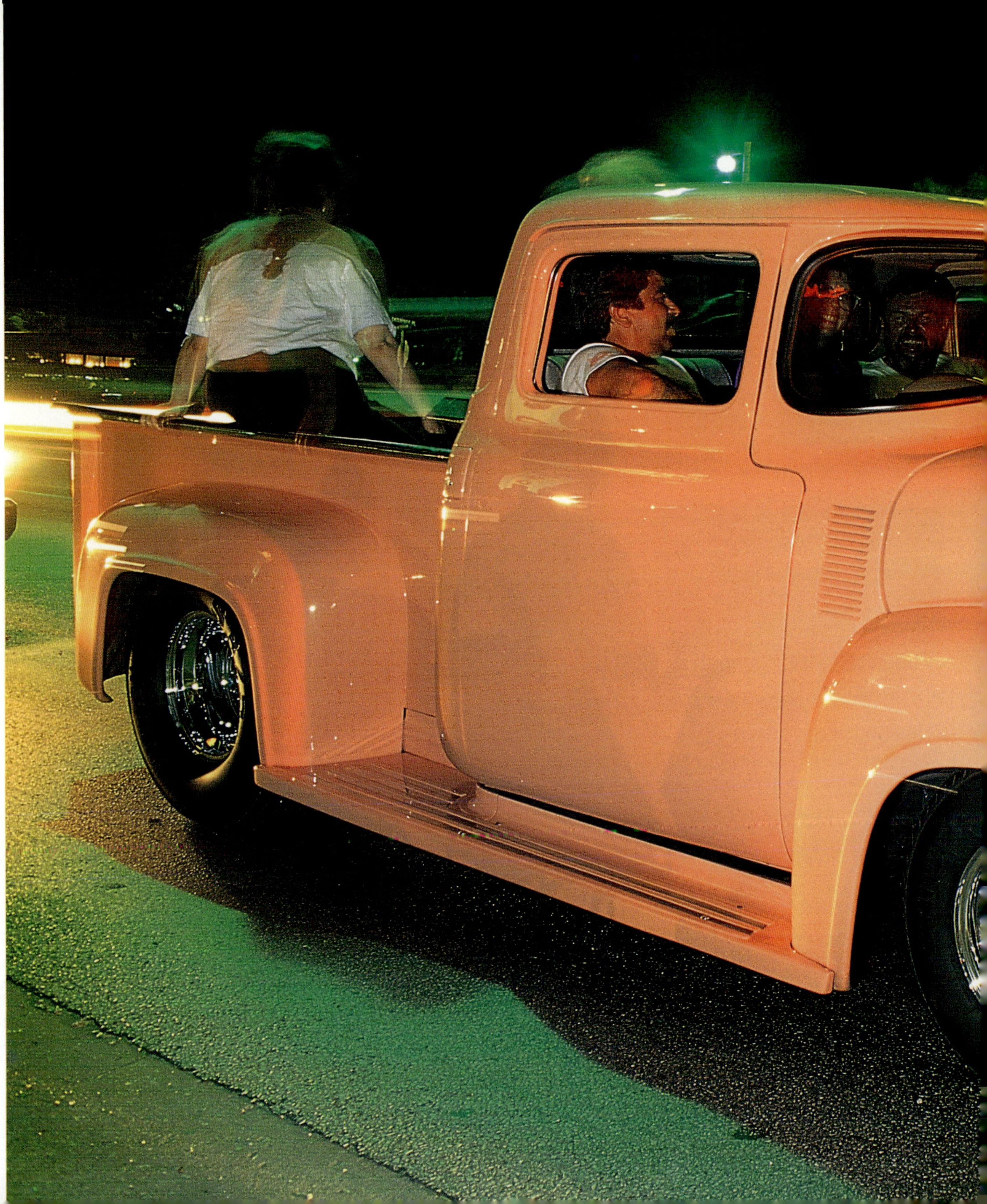

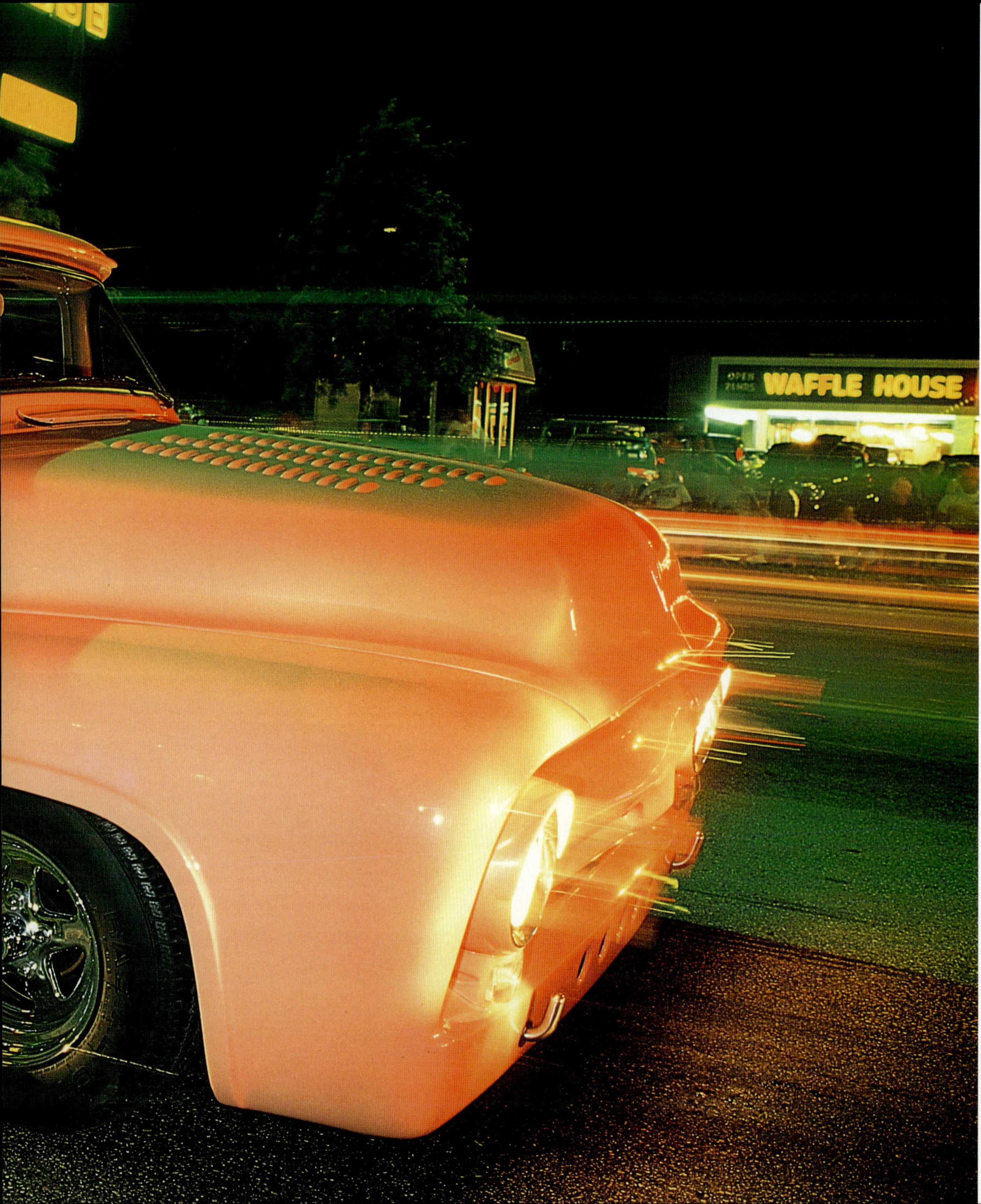
WAFFLE HOUSE

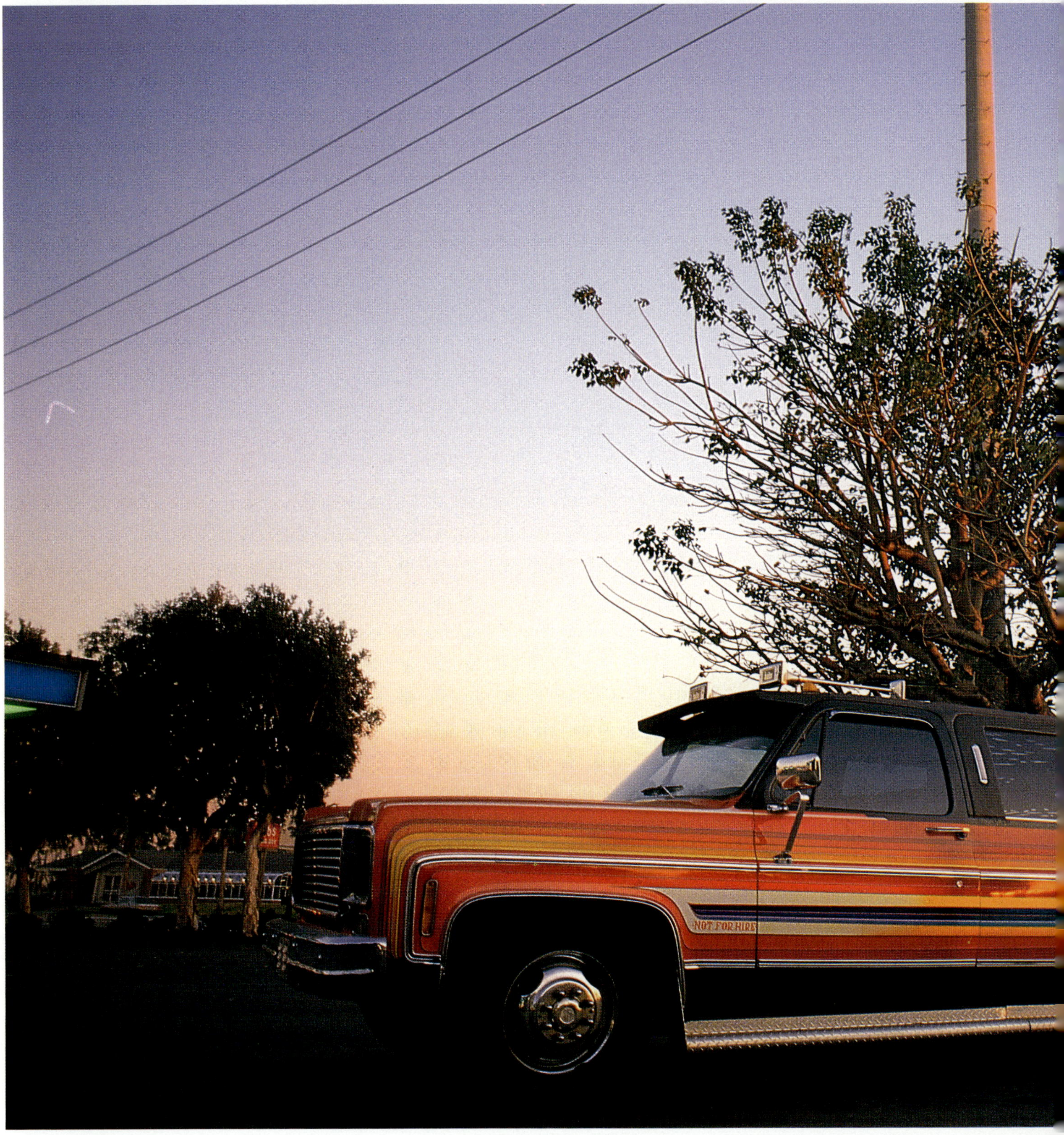
NOT FOR HIRE

Late-model Showcase

When it comes to customising more contemporary pick-ups, the imagination applied to a wide variety of machinery is simply amazing. Although most enthusiasts inevitably follow prevailing trends when deciding how to modify their trucks, there is far less discrimination against those who 'dare to be different' and come up with an alternative to the norm.

Perhaps due to the relatively conservative styling of post-1960 trucks compared to earlier models, or simply because they exist in larger numbers, owners of late-model pick-ups appear to have fewer qualms about making radical changes to their vehicles. Chopped tops are certainly not unusual, and most seem to lose most of their original exterior

Parked at a gas station on the Pacific Coast Highway at Huntington Beach, this limo truck was 'not for hire'. One can only assume it earns its keep as some kind of publicity vehicle or courtesy rig, possibly collecting rock bands from the nearby airport. Think of all the equipment it would carry

trim in the transformation from stocker to custom.

Paintwork, as ever, is largely determined by the fashion of the day. Not so long ago complex special effects were all the rage using Metalflake lacquer and a bewildering array of special finishes. By the mid-Eighties pastel colours were in, following the lead set by the 'Cal-look' VW enthusiasts, and colour-matched bumpers, grille and trim was the way to go. More recently, there has been a veritable explosion of vibrant colour in conjunction with 'new age' graphic designs.

High performance continues to excite many light truck owners and the old dictum – 'There's no substitute for cubic inches' – still holds true. Engines are beautifully detailed, being painted and polished to perfection and often adorned with the latest hi-tech milled aluminium anciliaries.

In contrast to the extreme ride height preferred by the four-wheel drive crowd, those limited to two-wheel drive usually go to great lengths to ensure their vehicles sit as low as possible. Smooth and slippery, minus door handles, trim, tie-down hooks, and so on, body mods are limited to wind-cheating aerofoils, flared wheelarches and rear-mounted race-style aerofoils. And for those who favour the 'Pro-street' look, the task of squeezing immense rear tyres inside the original body lines is somewhat easier done on a pick-up compared to a passenger car, though the effect is nonetheless impressive.

The ever-increasing number of Volkswagen enthusiasts are sadly limited in their choice when it comes to pick-ups. The early Sixties Type 2 models are not exactly plentiful today, but they make impressive customs given the 'Cal-look' treatment. This consists of radically lowered suspension and trick wheels, a bright paint-job and a suitably re-modelled interior. Alternatively, Beetle fans may purchase a glassfibre conversion kit to turn their *People's Car* into a Baja desert racing flat-bed pick-up, in contrast to the more traditional Baja Bug.

The introduction of small Japanese pick-ups to the US market in the Seventies led to

Take a 1977 Chevrolet Fleetside, lose about 6 in. from the roof pillars, shave the door handles, punch the hood with 160 louvres, custom paint with outrageous flames and what have you got? An original traffic-stopping street machine, that's what! The front of this pearl yellow and Candy orange hot-lick special has a tube grille for that trick custom look, the suspension has been lowered by cutting two coils from the front springs, and 454 cu. in. of Chevy V8 ensures it's always an odds-on contender in the traffic light grand prix

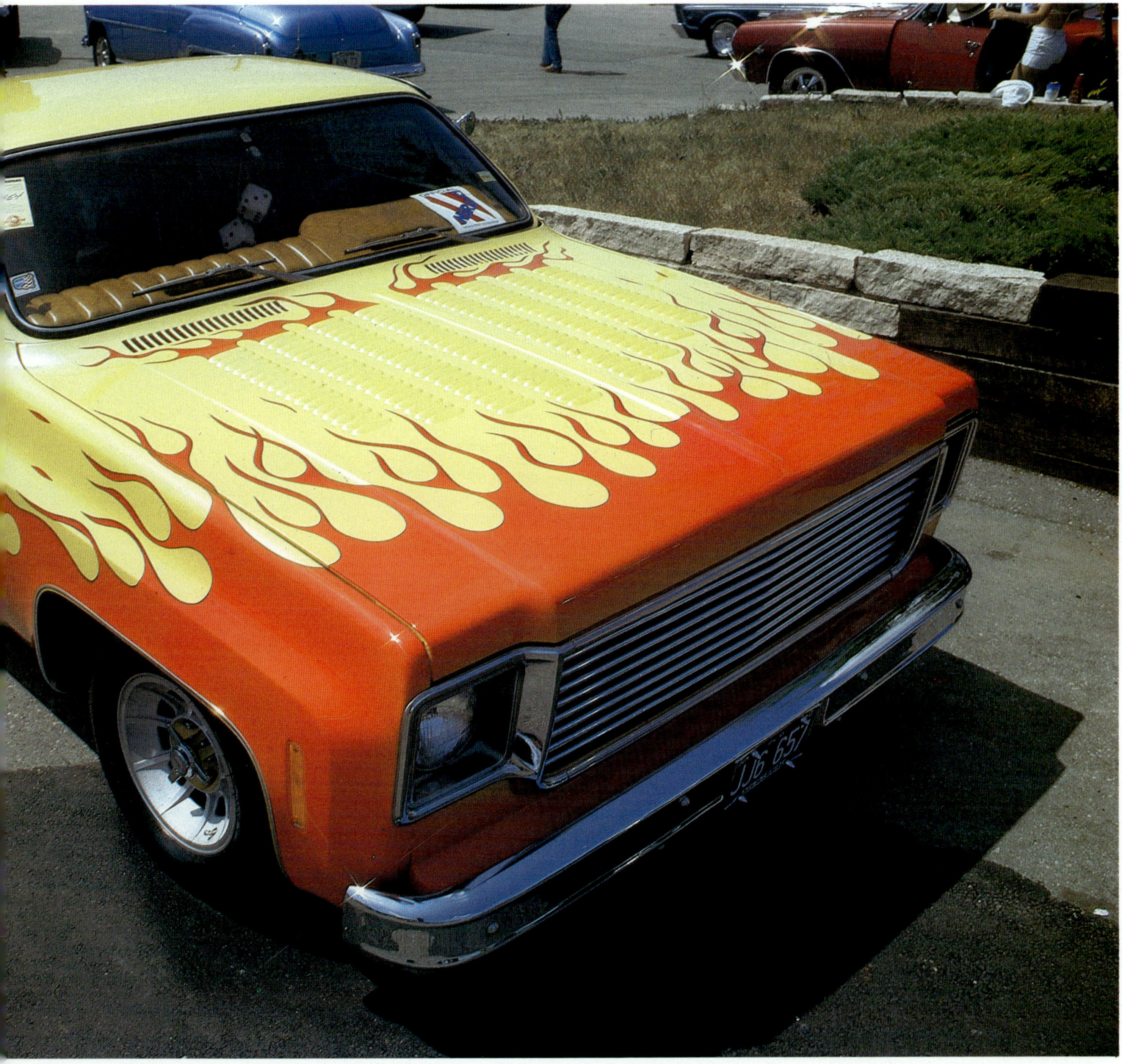

the current huge interest in modified mini-trucks. Naturally, the phenomenon started in Southern California where any teenager with a credit rating could suddenly afford a brand new truck. This situation was quickly exploited by the aftermarket custom accessory manufacturers and suddenly every kid on the block was lowering his 'mini' to the limit and fitting it with a custom grille, lights, and so on.

Today, there are vast numbers of these modified imports cruising Main Streets all over the USA and beyond. Sub-fads have arisen within the scene and it has become a contest in numerous clubs to see who can come up with a more radical rendition than the last. Indeed mini-trucks, like pick-ups in general, are hotter than ever!

Above
Subtle graphics on this low-buck, low-down, low-lid Chevy make for an impressive statement. Late Sixties or early Seventies trucks are the way to go for those on a budget — low on cost, high on street cred'

Right
A mural that any Chicano would be proud of. Airbrush and paintbrush work like this may have been replaced by bolder, brasher graphics, but this scene of a delicate footbridge over a tall waterfall is enough to make any trucker go all sentimental. Isn't it?

Left
Alternative tailgate treatment features inverted louvres and clever graphics. The blue dot on the rear light is just that: a 'Blue Dot'. Beloved by hot rodders and hated by cops everywhere, this little glass lens accessory glows a purpley blue at night and for some reason has been one of the most popular add-ons for decades. Why the appeal? Who knows, but no real hot rod is without them

81

One from the archives. Custom paintwork of this nature was very popular in the Seventies. Murals depicting the darker fantasies of ghouls, ghosts and graveyards were especially popular, and needless to say, a great deal of skill was required to execute a job of this artistry and complexity. There are extensive body mods to this '76 Chevy as well. The inset rear lights are obvious, but the trick tailgate and hammered roofline are also well done. They don't build 'em like they used to, eh

What a brute! With its radical roof chop, custom bodywork and big-block V8 rumble, this LO STEP *turns heads wherever it goes. In this case, the* Car Craft Magazine *Street Machine Nationals in Springfield, Illinois*

83

Is it a car or is it a pick-up? The answer is yes, of course. Chevrolet's El Camino and Ford's similar Ranchero model combine the best of both worlds: the carrying capacity of a small truck with the looks and comfort of a passenger car. They have become so popular in America since their introduction in the late Fifties that there are even sports versions with high performance engine options. No matter that these dual-purpose vehicles are rarely subjected to the rigours of a hard working life, there is plenty of room for two and you are never short of space for the odd six-pack. This '72 El Camino is rather showing its age, however. Although expertly painted and beautifully upholstered, this rather garish look is now generally considered passé

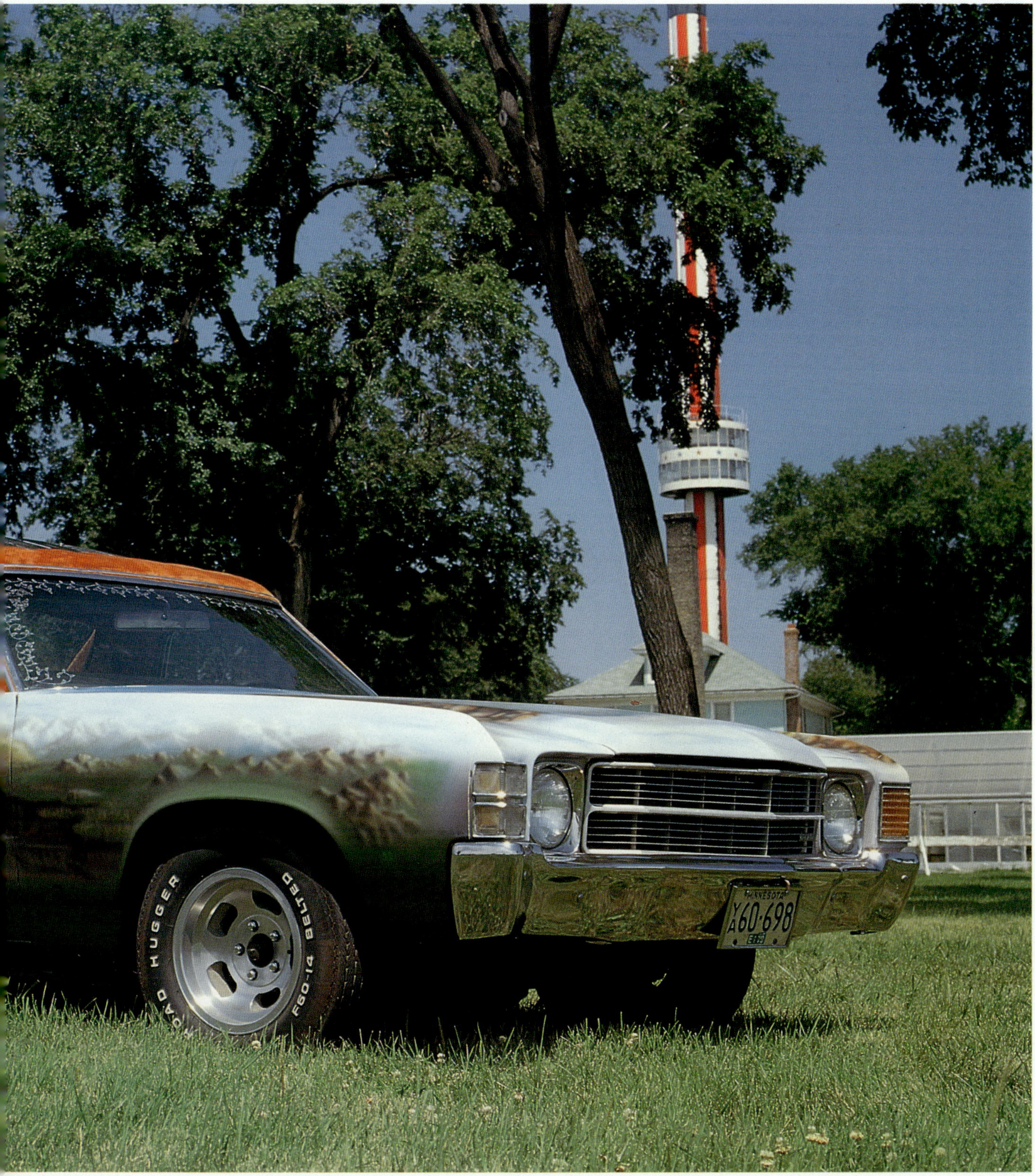
HUGGER
F60-14 BELTED
60·698

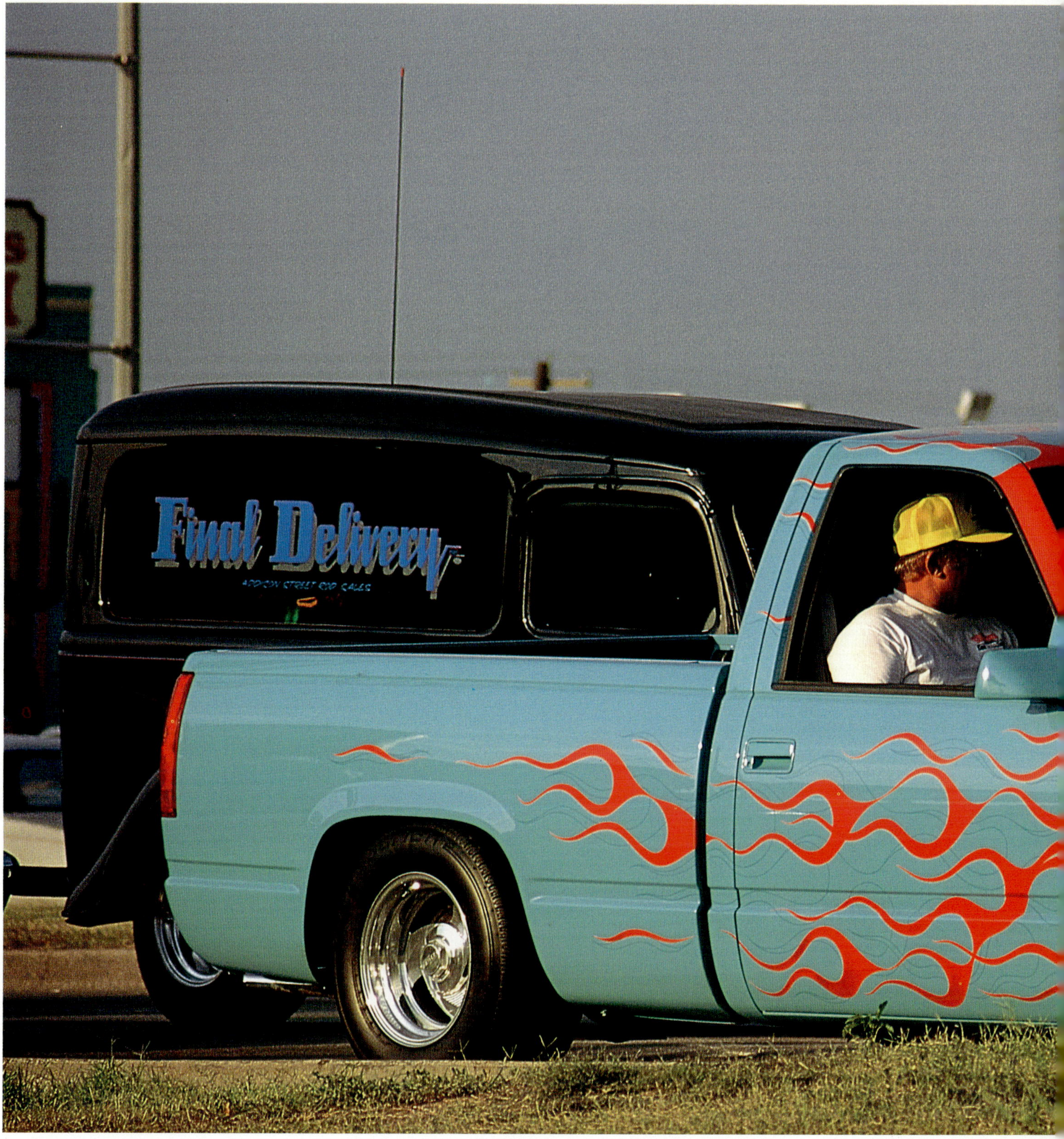
Final Delivery

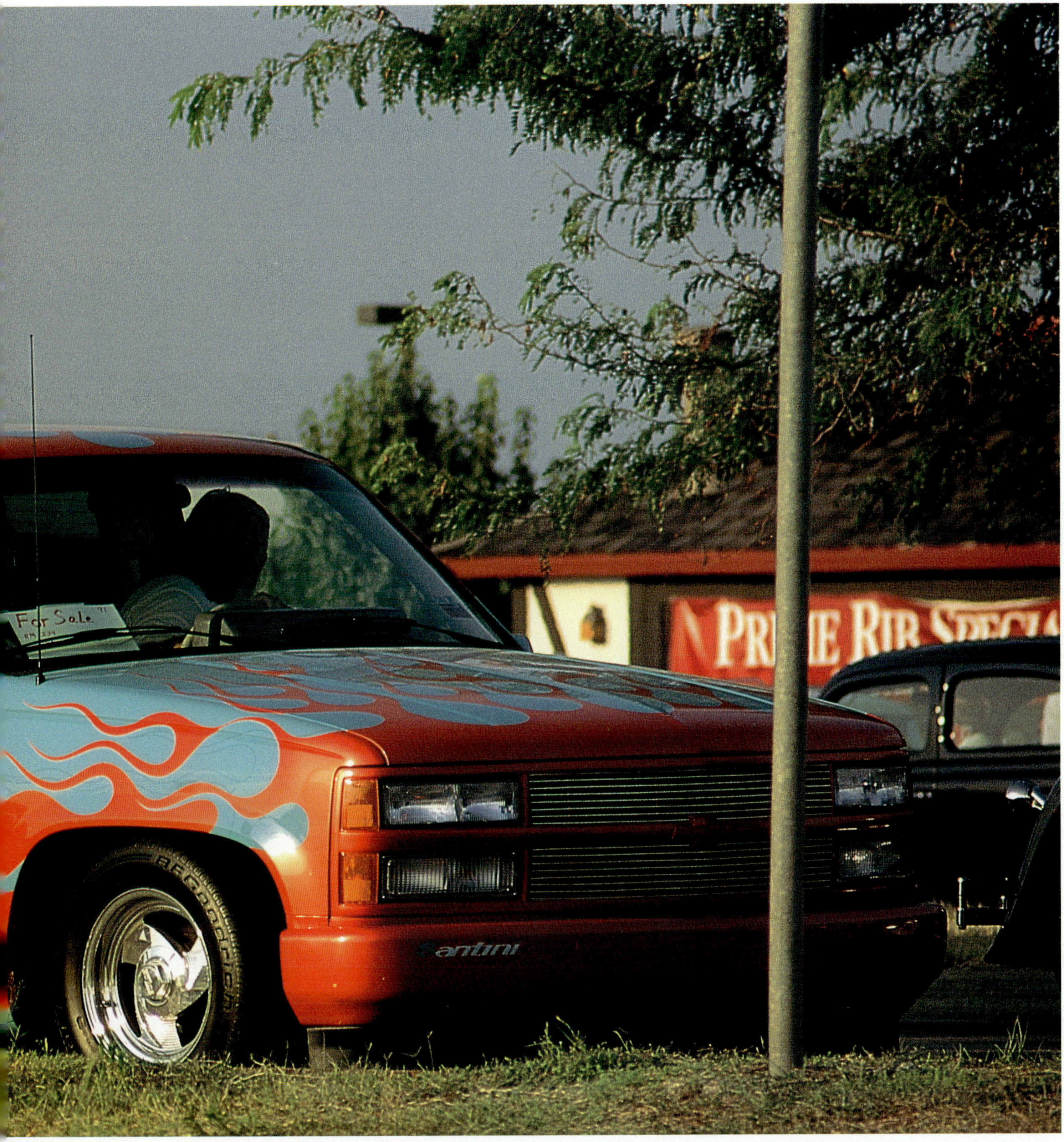
For Sale

Previous page
Late-model Chevy Fleetside demonstrates the basic art of hot rodding. A lowering kit, custom wheels and Candy flames is all it takes

Above
Just look at the width of the rear wheels on this GMC Sierra Classic 'Wide Body' then compare them to the fronts. This combination is 'Pro-street', with a capital 'P'! Apart from the weld rims and some rather subtle pinstriping, this truck looks pretty well stock. But there must be some serious 'tubs under the tarp' to cover all that rubber and, no doubt, a hot one in that capacious engine bay

Above
In the colourful world of custom pick-ups, a two-tone paint scheme can mean anything you like. Deep Centerline wheels, on the other hand, convey a more specific message...

Left
Check the trick, airbrushed 'Chevrolet' on this '75 model. Bitchin'!

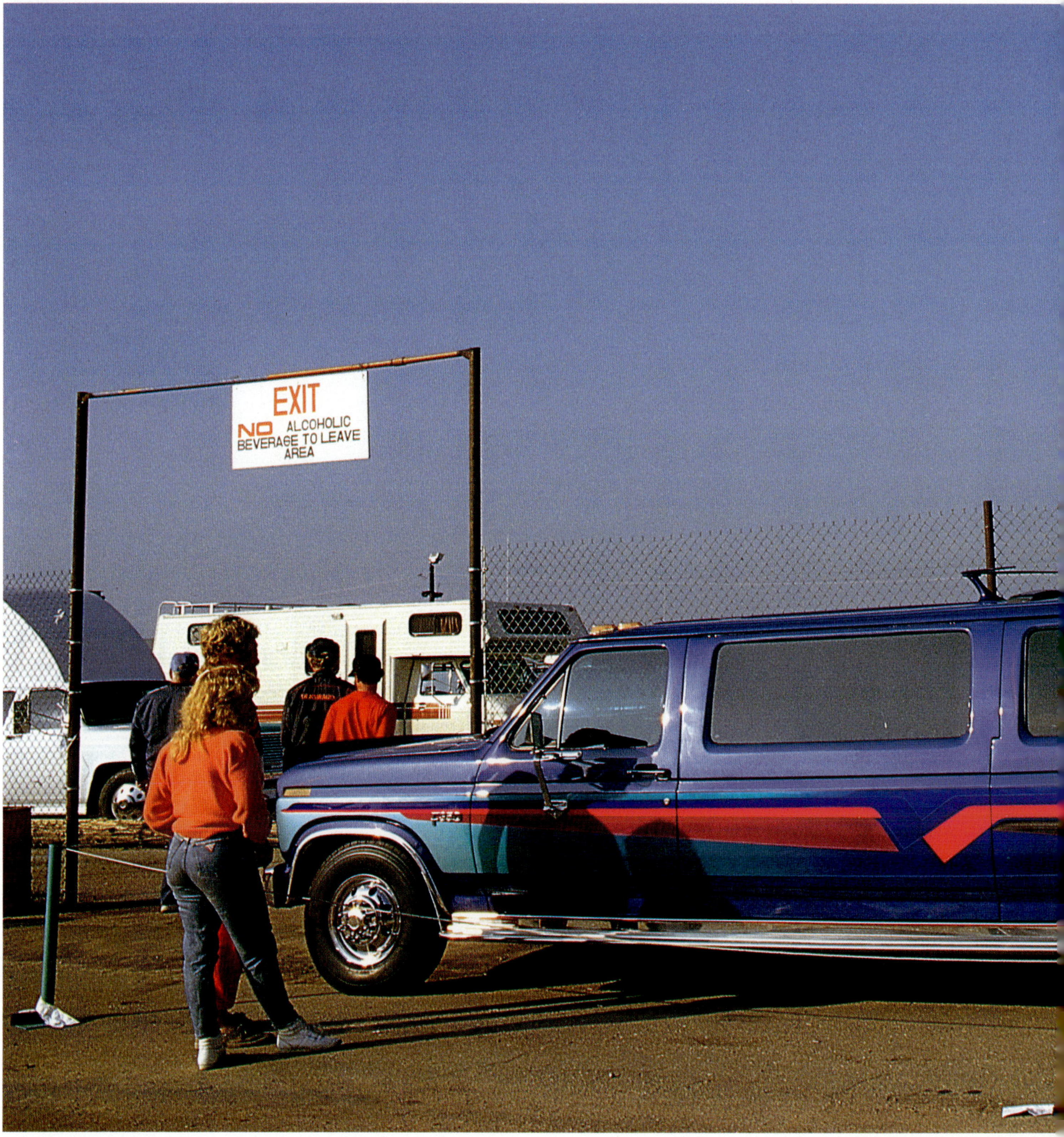
EXIT
NO ALCOHOLIC
BEVERAGE TO LEAVE
AREA

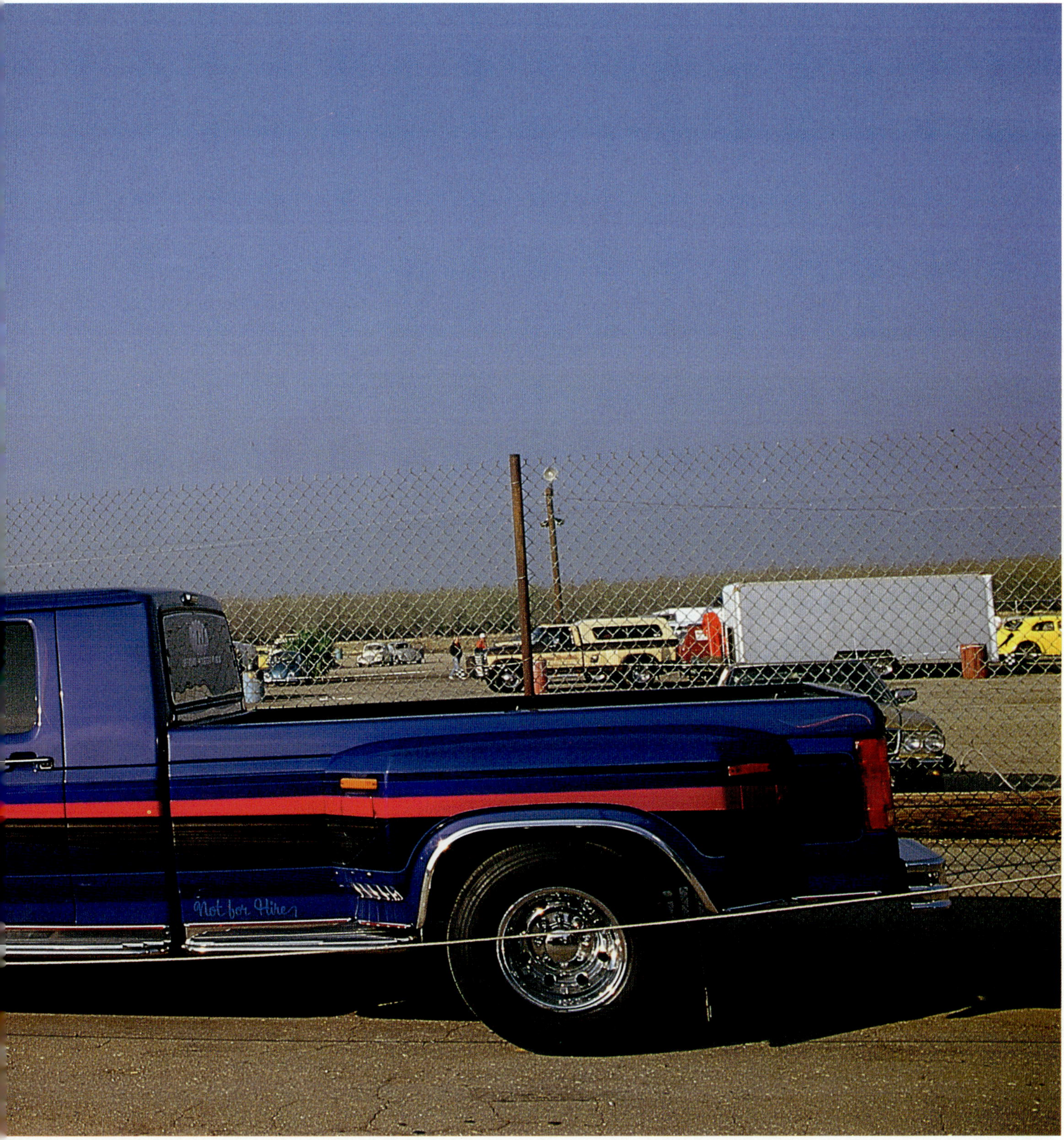
Not for Hire

Previous page
A stretch limo is one thing, a stretch truck is something else. Ford's F.350 may be ordered from the factory with a four-door cab to carry a larger crew. This custom-built job can not only accomodate the crew but the entire boardroom too!

Left
Stretched Rabbit: Cleverly converted Volkswagen incorporates the rear doors from a car between the original cab and pick-up bed – or was the Rabbit bed grafted onto the rear of a 4-door car? No matter. The combination of the two is unique, resulting in a really useful vehicle with room for 4–5 passengers and a huge load capacity for its size. It is the El Camino/Ranchero concept taken one step further

Below Left
For some reason Volkswagen never produced a Beetle pick-up. However, several companies have manufactured DIY kits over the years which capitalise on the inherent strength of the VW chassis. This unique Bug is expertly finished with a hand-crafted stake-bed, a Baja racing-style hood and over-size replacement fenders to cover the polished alloy wheels

Right
Early crew-cab VW pick-ups are considered 'cool' within the Cal-look VW scene, and Rob Anderson's lowered and louvred KOOL CAB *must be the coolest around. The 1960 model features umpteen minor mods, none of which detract from the truck's inherent character*

Below right
Hey, shorty! Does your mother know you're out? Severely truncated VW Buses and pick-ups were all the rage at one time, but they are rarely seen today. Around 3 ft has been removed from the centre of this 1960's split-screen, and the combination of rear-mounted engine, lightweight front end and short wheelbase enables the driver to pop 'wheelies' at the drop of a clutch. Whacky!

93

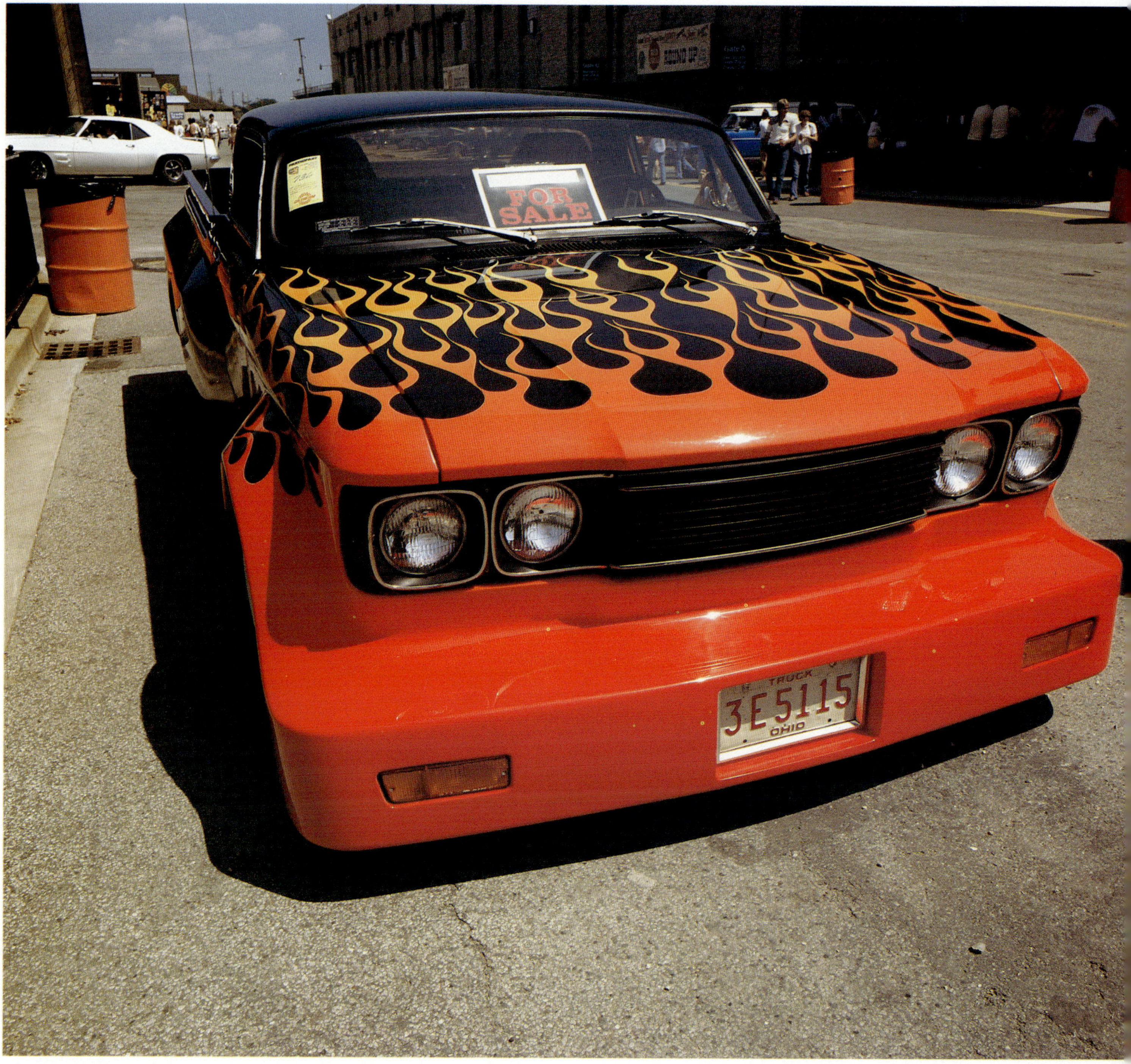
FOR SALE
TRUCK
3E 5115
OHIO

Left
Too hot! Used Chevy LUVs and similar light trucks are cheap to buy and quite easy to modify, in as much as they are supported by a large aftermarket industry. Then again, you may prefer to buy one ready-built...

Above
In the fight for originality, imaginations have exploded with colour. Witness this 'shattering' effect on a late-model chopped cruiser

Overleaf
Angelo's is a well-known hangout for hot rodders of all persuasions, situated close to Disneyland in Southern California. There you will find, on selected nights, all manner of street machinery including plenty of hot mini-trucks like this pink roadster. In contrast to the large number of trucks with cut-down windshields, this one retains the original factory glass

95

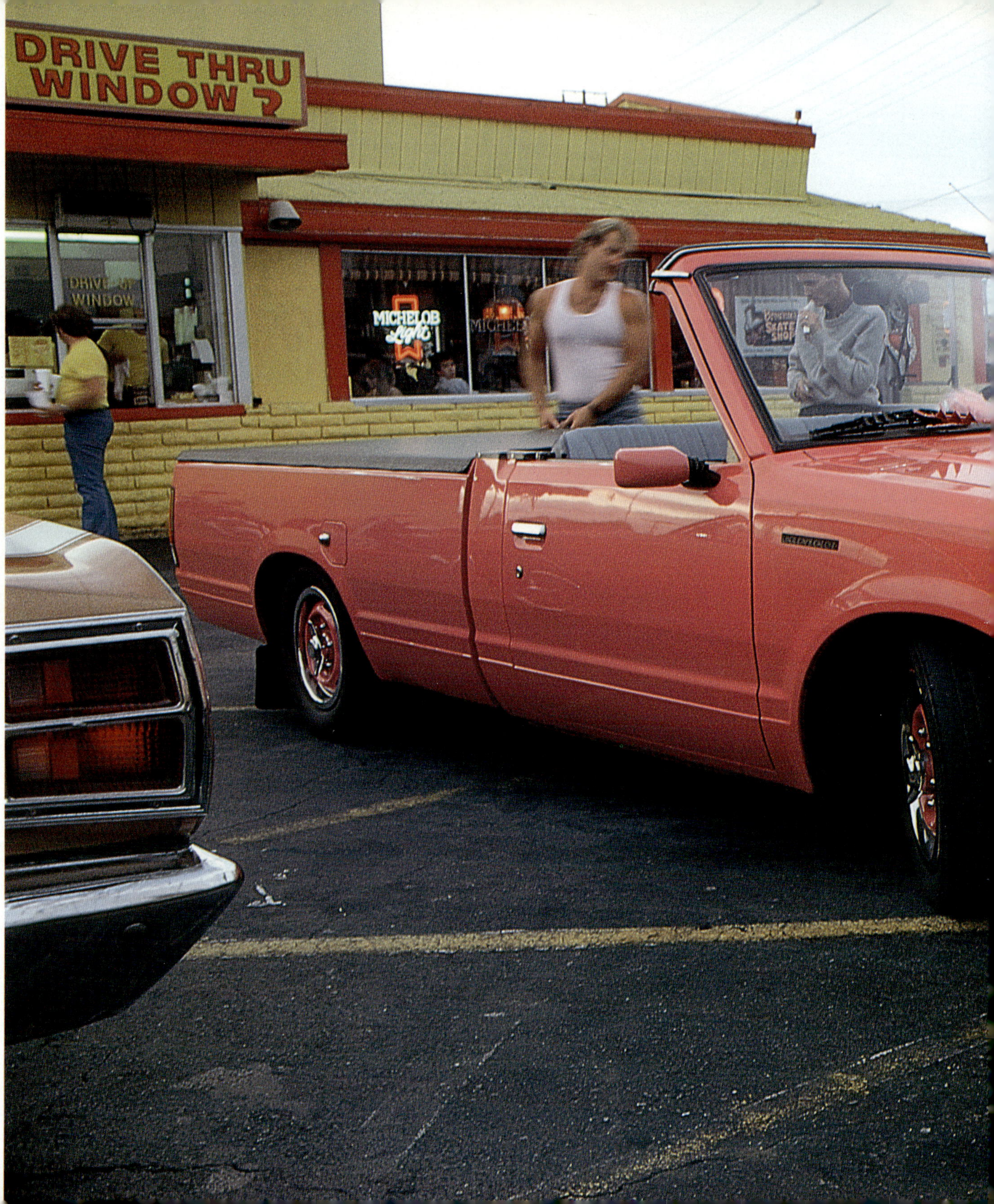
DRIVE THRU
WINDOW ?
MICHELOB
Light

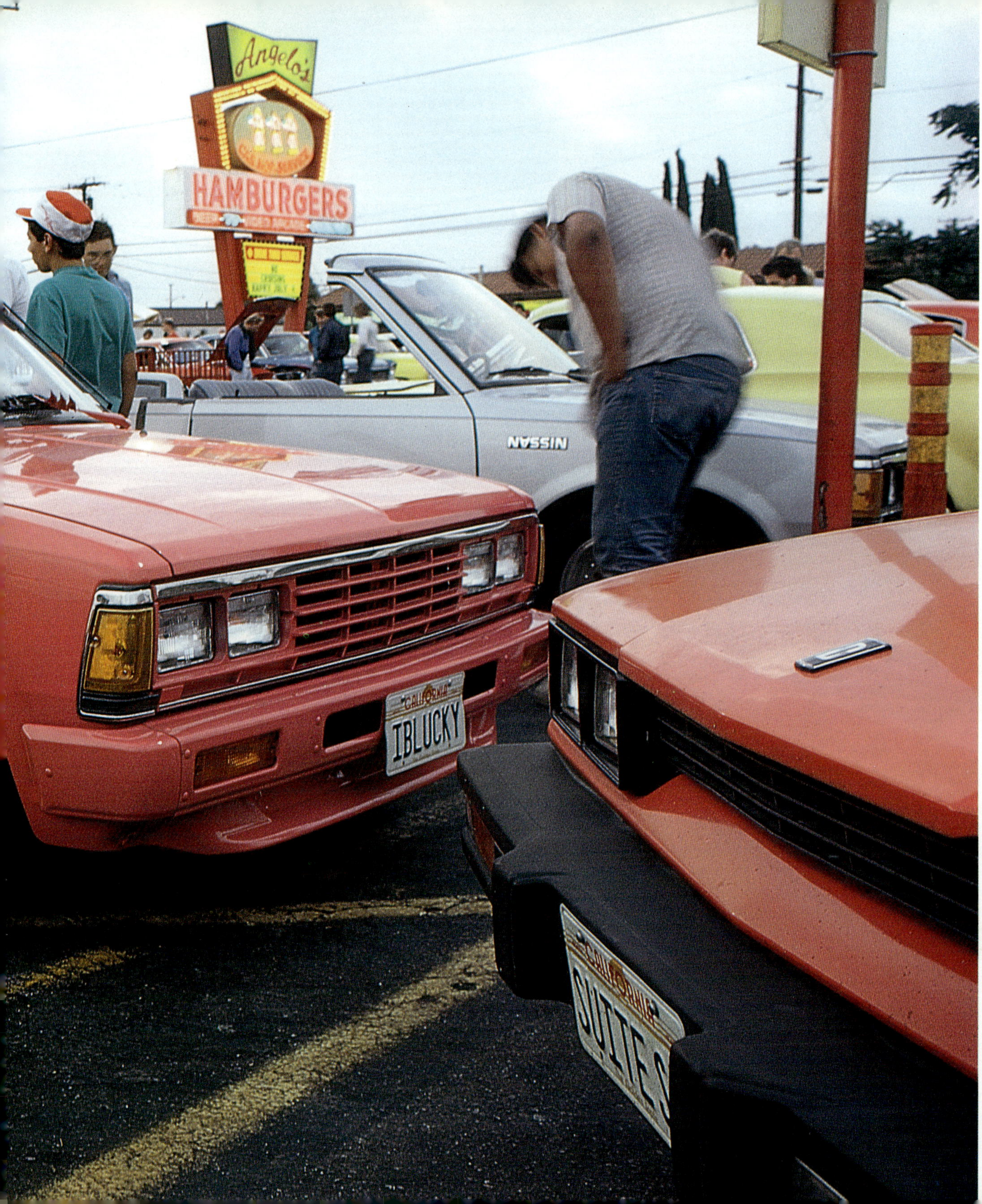
Angelo's
HAMBURGERS
IBLUCKY

Super-clean Mazda showing off its seriously protruding but oh-so fashionable wheel and tyre combination in Los Angeles. Skinny low-profile radials are stretched over wide, mirror-finish rims to achieve the 'rubber-band' look; a typically quirky LA trend which, it has to be said, is decidedly suspect in terms of safety

Left
Early Japanese 'mini' with flared wheelarches and graphic paint dates from the early Eighties — impressive, nonetheless

Below
Mazda B2200 demonstrates the more restrained side of California mini-trucking, avoiding brash graphics in favour of a plain Jane but ever-so clean metallic paint-job. Needless to say, the truck has been 'slammed' (lowered) to the limit

One of the trendiest mini-trucks in the US a few years ago. Based on a 1978 Datsun, this pro-built machine was up-rated in every department. The convertible top conversion and IMSA-styled circuit racing fender flare and air-dam package are fairly obvious changes. The windshield pillars were sectioned by 4 in. and the bed shortened by 12 in. The hood has been louvred and the tailgate smoothed over, while a Stull tubular grille with quad lights was substituted for the original arrangement and a hefty roll bar welded-in for safety. The truck is custom painted in pearl yellow with Candy bronze and copper stripes and the uprated suspension fitted with Compomotive modular wheels wrapped in BF Goodrich rubber. The stock motor features competition internals, a Weber carb and a Turbosonic turbocharger. All told, a 'rice-rocket' with show and go!

Chris Fehring installed a mid-mounted 455 cu. in. Oldsmobile lump in his open-top Chevy LUV in the quest for originality and performance. Liberal use of brown velvet material has certainly created plush surroundings for it — let's hope there are no oil leaks

Four-Wheel Jive

In recent years, the popularity of four-wheel drive vehicles has escalated beyond all proportion. There are several reasons why so many have invested in these recreational vehicles, not least the appeal of the great outdoors, though quite a high proportion are rarely, if ever, driven off paved roads. You see them cruising the boulevards; you see them at the grocery store; you see them at car shows, polished to perfection with hardly a speck of dirt showing.

The typical 'four-by' pick-up, a Chevy, Ford, Dodge or GMC, comes from the manufacturer with a mid-size V8 engine and automatic transmission backed up by a transfer case which allows manual shifting from two-wheel to four-wheel drive. The front axle has a differential and, to provide clearance from the chassis, both front and rear axles

Right and overleaf
Imagine the reaction to this beast on the streets of your home town. A cut-above the average boonie-basher, that's for sure!

102

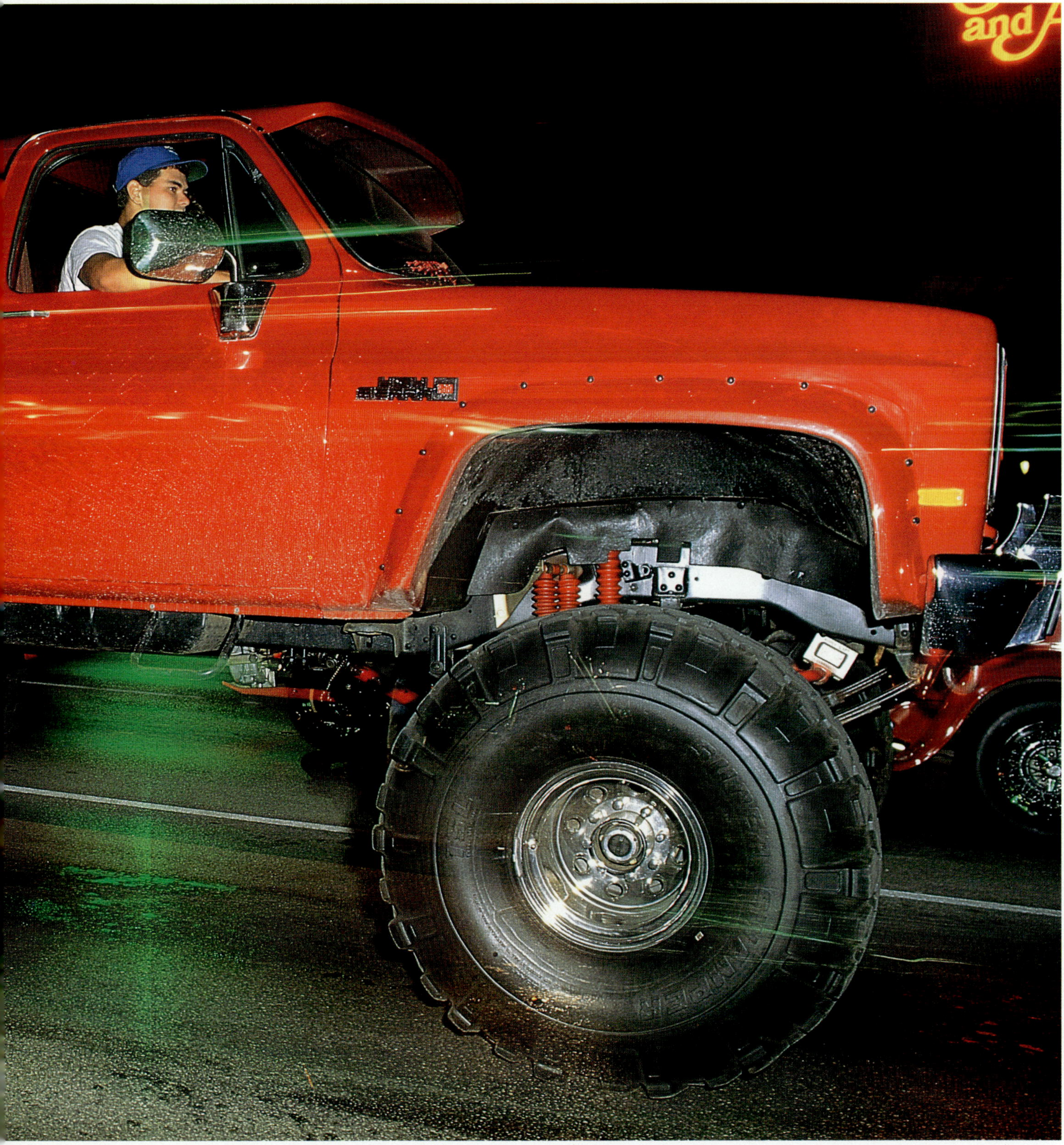
and

are mounted below the springs, thus raising the entire truck at least two feet off the ground.

What separates the average custom 4 × 4 from the factory models and puts it well and truly into the posing class are the many extras which have become hallmarks. Most important is a set of super-wide chromed wheels shod with behemoth 'all-terrain' tyres. Also imperative is a higher-than-normal ride height and a heavy-duty chromed roll bar, topped with at least a couple of high-intensity driving lights. Protecting (or dressing-up) the front end is a brush guard, and the more serious will add skid plates under the front differential. Naturally, a custom paint-job and a mega-watt sound system are considered *de rigueur*. The typical custom 4 × 4 gets driven to school, work or around town during the week, then to the local cruise venue at weekends. The calculated effect is outrageous.

But not as outrageous as the sight of a genuine 'monster-truck' flying through the air and crashing down onto piles of automobiles! The monster-truck fad began in the mid-Seventies when Missouri's Bob Chandler came up with a novel idea to promote his *Midwest Four Wheel Drive Centre*. 'Bigfoot', a blue Ford pick-up, featured gargantuan tyres and a

104

supercharged engine and demonstrated its might at county fairs by driving up to piles of junked cars, shifting into low gear and simply driving over everything in its path!

Bigfoot begat bigger Bigfoots which started attracting competition. By the early Eighties, monster-truck exhibitions had become regular attractions at various events and the trucks got fancier, growing as high as 15 ft, featuring four-wheel steering and flame-thrower exhaust pipes and costing $150,000. Today, enthusiasts fill fairground tracks and indoor arenas to watch sanctioned monster-truck racing and car-crushing action at speeds up to 50 mph, and the sport is seen on TV screens all over the world. It could only have started in America...

Above
Likely lads of Springfield, Illinois looking for the action in a late Seventies Ford Bronco Ranger XLT. The XLT differed from the base-model Bronco in many ways, featuring rectangular headlamps, additional brightwork and various convenience options. The green plastic bug deflector, smoked Perspex headlight covers and roof-mounted marker lights were fitted by the owner

106

Left
Tough-looking Chevrolet Silverado was photographed at the Street Machine Nationals in Milwaukee way back in 1979, but there are hundreds, if not thousands just like it on American streets today

Above
'Sky High', a much-beautified early Eighties Ford Bronco, provides a suitable resting place beneath its chassis for the one 'shade-worshipper' in this group

Overleaf
Boulevard bruiser: Riding high on the streets of Illinois, this late Seventies Ford has probably never been off paved roads in its life. The crowds were urging its driver to 'lay rubber'...

AMERICAN & ITA
THE MAIN GAT
BUSCH
ON TAP
THE
MAIN GATE
Enjoy
Coca-Cola
Enjo
Coca

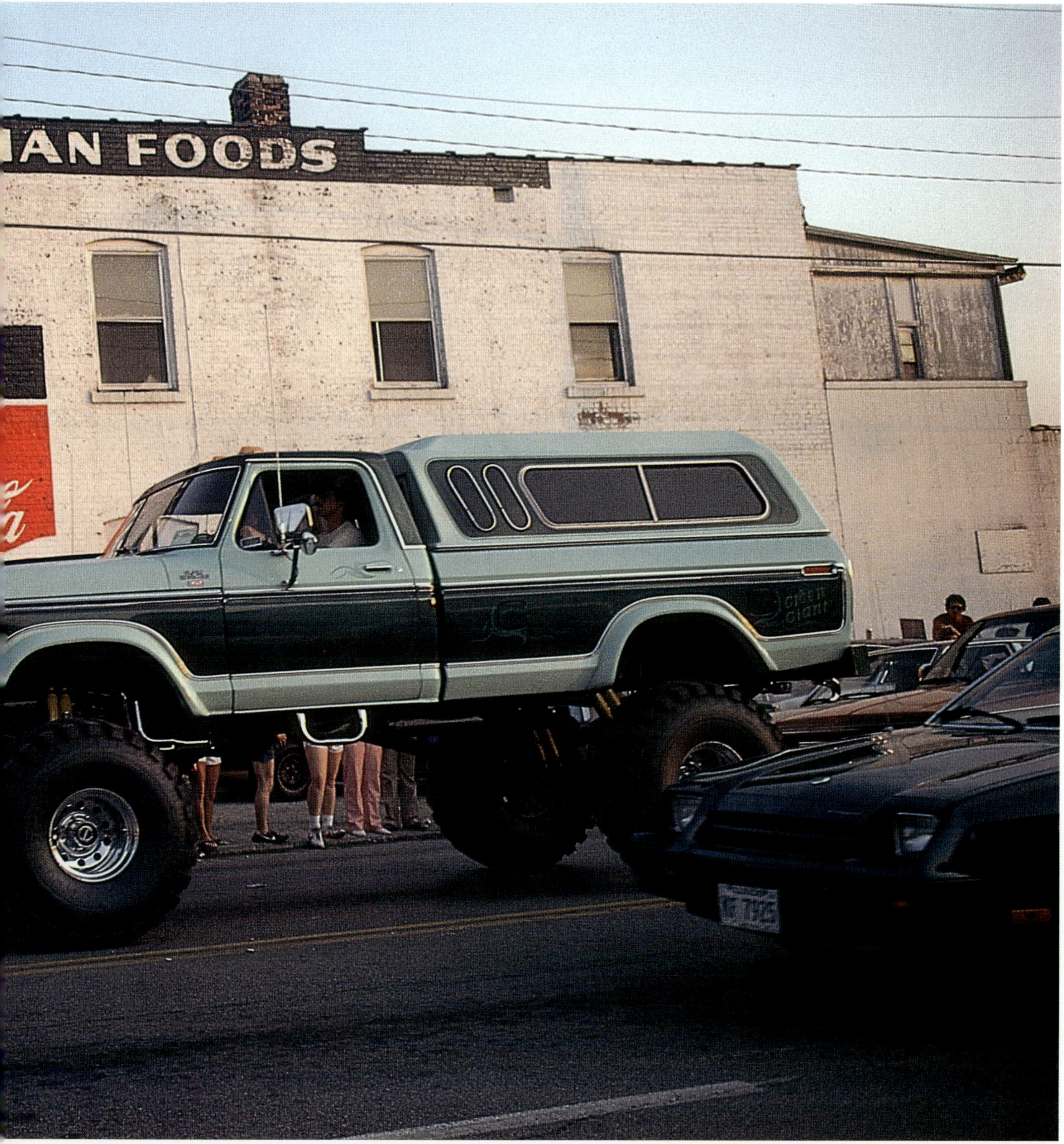
AN FOODS
Green Giant

Customised Ford F.350 Ranger 'Styleside' (Ford parlance for 'Fleetside') has at least 13 more lights than originally supplied and claims to be 'Second Ta None'

The custom grille on this mid-Eighties full-size Chevy Fleetside was not a factory option. Ditto the mini camper-cum-storage box behind the cab

'Four-by' with a difference. The body of this '68 Chevrolet El Camino appears to be factory stock apart from the sunroof, but what happened to the running gear? While not immensely popular, the placing of car and light truck bodies onto 4 × 4 chassis' is not too uncommon...

Left
Full-tilt vintage truck has been the subject of much ingenuity. Bet it's never been off-road though

Below
One-of-a-kind 1954 Ford F.100 stepside. Four wheel drive conversions such as this are relatively cheap to perform using parts from the wrecking yard

Quite possibly the only customised four-wheel drive '49–53 Studebaker truck on earth. You've got to admire the owner's style...

While street rodders and mini-truckers strive to get their rides as low as possible, monster-truck enthusiasts reach for the clouds. This Dodge Power Wagon was once a regular civilianised version of the ¾ ton truck used by the US Military during World War II. Tall enough, eh

Everybody has seen or at least heard of 'Bigfoot', the first, and for many, the ultimate 'monster-truck'. Numerous articles have been written about the truck and its creator, Bob Chandler, ever since the first vehicle bearing the name was unleashed many years ago, performing in stadiums all across the USA. Car-crushing has since become a major spectator sport along with tractor-pulling and mud-bog racing, and there are now large numbers of monster-trucks on the show and race circuit. But Chandler, always the innovator, has not rested on his laurels, unveiling new Bigfoots (Bigfeet?) every few years. Bigfoot III, seen here, was based on a '76 Ford F.250 chassis with new – at the time – 1984 bodywork. Powered by a supercharged 460 cu. in. V8 mated to a Ford C-6 transmission, the truck's 5-ton military axles are individually steered. The front wheels are controlled by the steering wheel while the rears are hydraulically operated via a dash-mounted switch. And, of course, it is the wheels and tyres everybody notices – eight of them on occasion! Each Goodyear Terra tyre is 66 in. tall and when doubled-up for eight-wheeled car-crushing action, Bigfoot is an amazing 20 ft wide and even capable of floating in water!

Left
With Bigfoot receiving sponsorship and acclaim for Ford, it was inevitable that a General Motors-based vehicle should come along and challenge their supremacy. Witness 'Bearfoot', based on a Chevrolet pick-up and featuring all-wheel steering. This set-up facilitates extremely tight turns — useful in the confines of a small stadium or showground — and rather spectacular sideways manoeuvres

Below
With four-wheel drive and four-wheel steering, Bob Chandler's amazing 4 × 4 × 4 attracts massive crowds wherever it goes

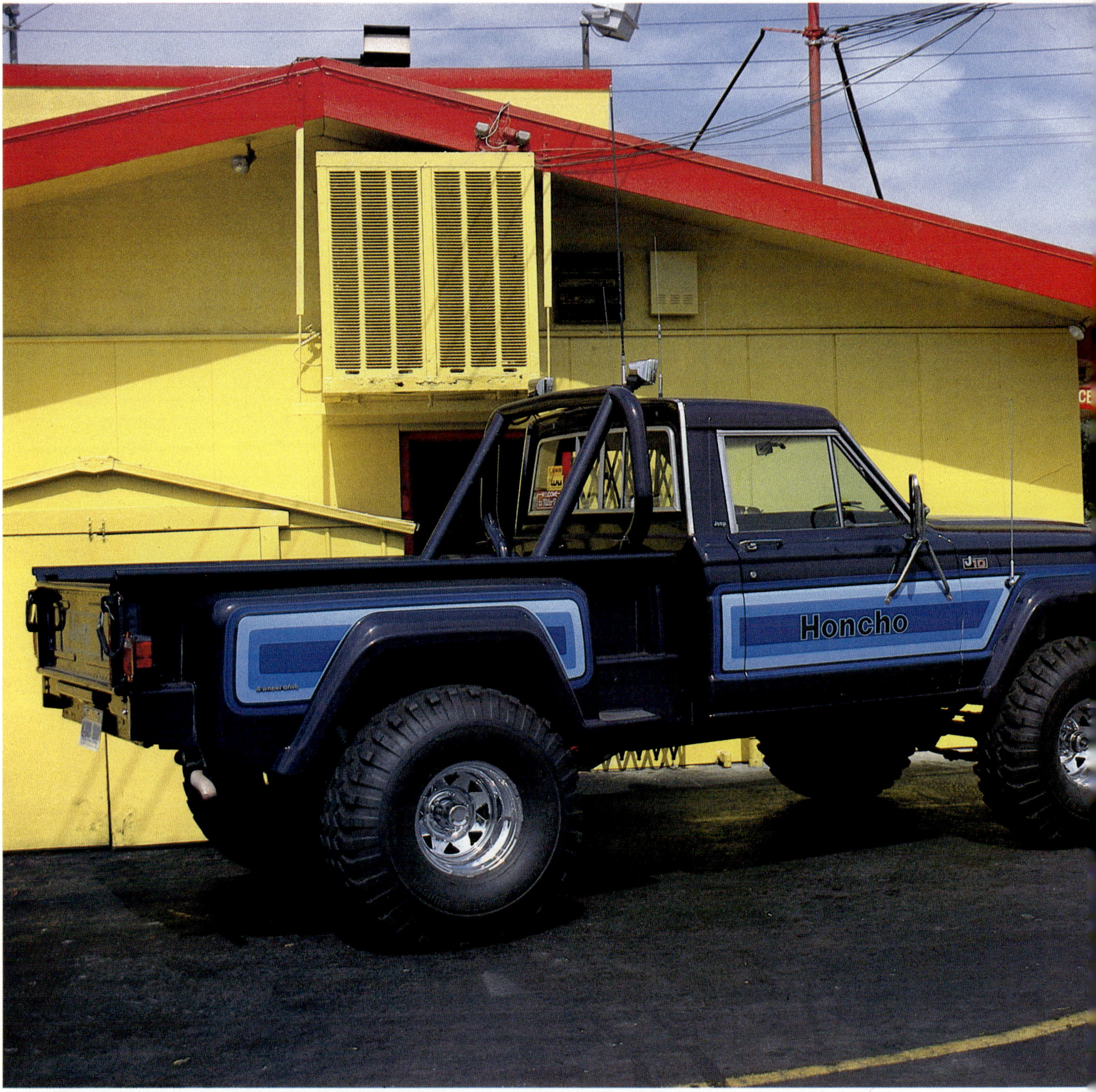
Honcho

Above
They say anything's possible in the Golden State...

Left
Super-clean Jeep Honcho poses outside Angelo's in Southern California. A suspension lift kit, custom wheels and chunky tyres is sufficient to transform a standard pick-up into a tasty street (credible) machine

Overleaf
Ford F.350 XLT 'duallie' is relatively stock. The bigger brother to the F.250 comes with optional 4-wheel drive and 5.8 litre V8, an 8 ft 'Styleside' bed and a choice of regular or crew cab. The latter, with four doors, seats six truckers in comfort

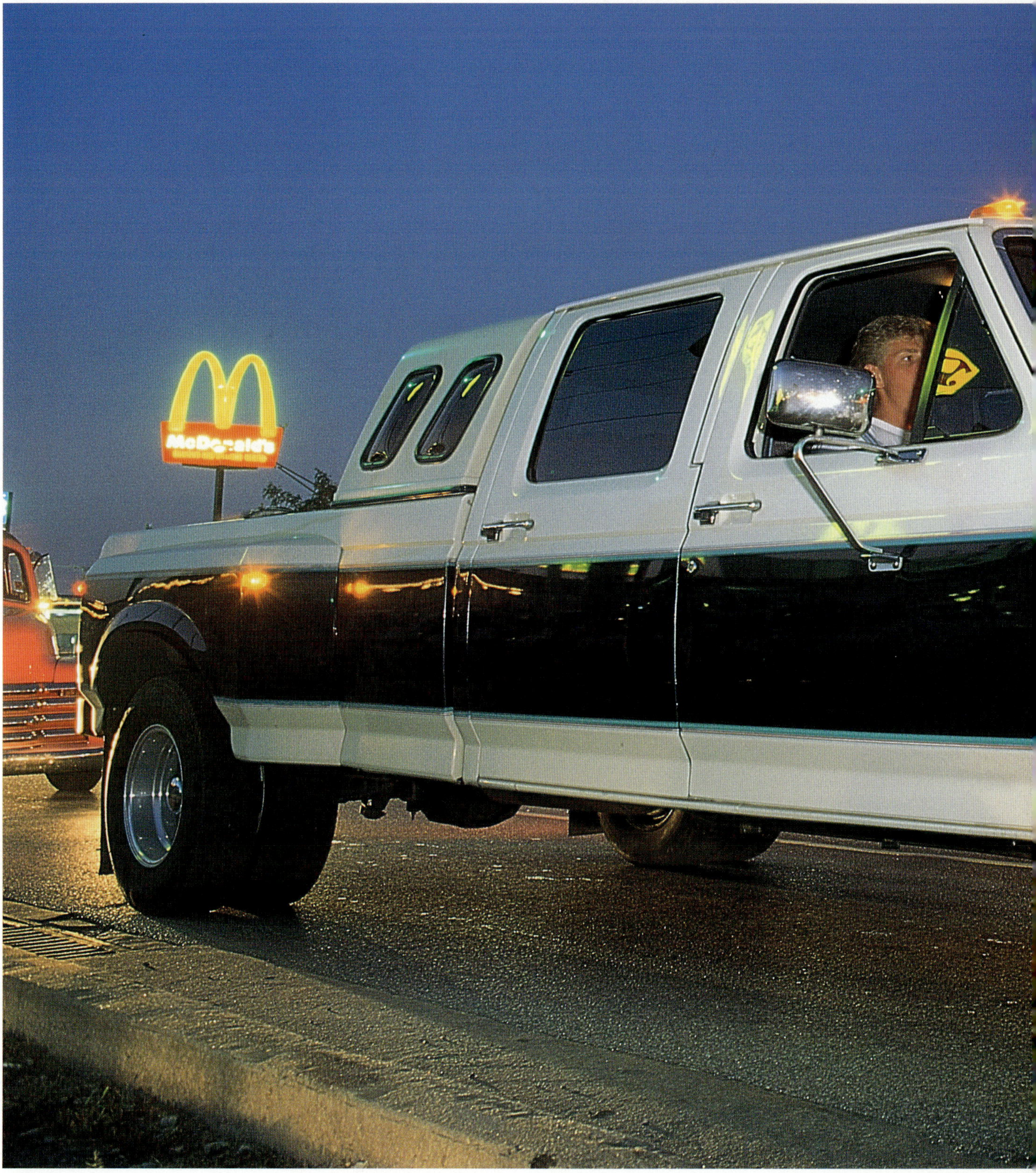
McDonald's

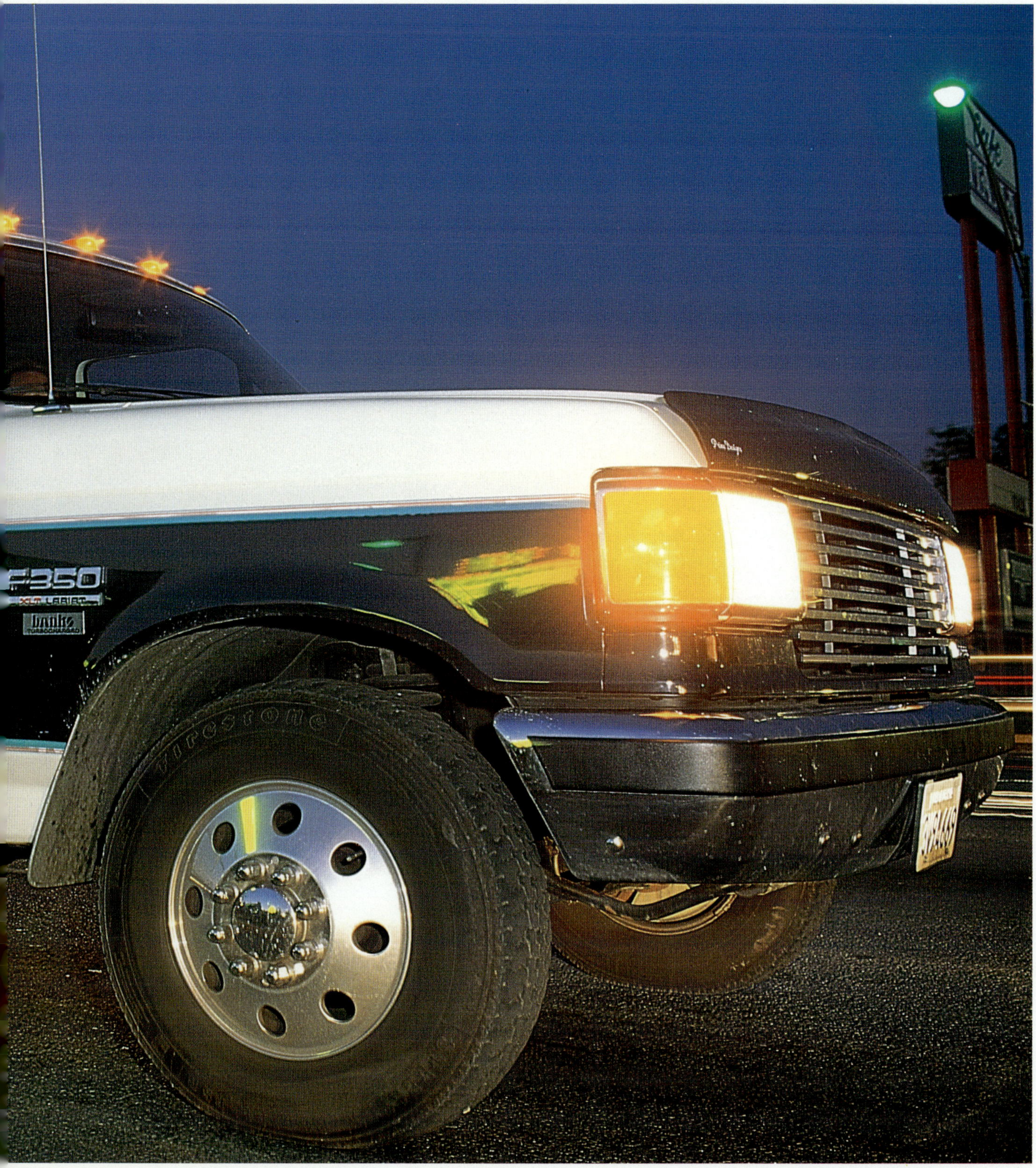
F350

One sure-fire way to win prizes at truck shows is to expose all your chassis detailing by way of a hydraulically-operated tilt bed

Left
A hot flame job and a set of polished modular wheels guarantees plenty of looks for the owner of this Toyota

Below
Reckon you would suffer from 'High Anxiety' driving this Toyota in high winds!

Creative Evolution
FUEL
SUN-MOON
DATSUN
18707 STUDEBAKER
CERRITOS, CA 90701
Firestone
ALL TRACTION UTILITY
21L-24

Gina
NISSAN
4 wheel drive
Firestone
21L-24
ALL TRACTION UTILITY
SHOCK FORTIFIED

Previous page
'Creative Evolution' is billed as 'The Biggest Datsun In the World – Bar None'. Quite a claim, but with good reason, because this particular 4 × 4 × 4 is a proper mini monster-truck. Starting life as a stock Datsun 4 × 4 in Redondo Beach, California, owner Ed Socie made the transformation by way of some rather radical mods to the running gear...

Left
57 in. tall Firestone tyres ride on suitably narrowed 2½ ton Rockwell military axles supported by custom-made leaf springs and no less than 16 KYB shock absorbers. This raises the overall height of the truck to an amazing 9 ft 7 in. Front end steering is via a conventional steering box from a Ford F.250, while the rear is steered with the aid of an electrically-operated hydraulic pump

Right
Power comes from a 1972 Chevrolet 350 cu. in. V8. Performance parts from Carter, Edelbrock, Sig Erson and Hooker assist in squeezing almost 400 bhp from the engine. The power is channeled through a beefed-up Turbo Hydro automatic transmission to a heavy-duty transfer case

Below left
Baby monster retains its California licence plate but is rarely driven on the street – thank goodness!

High-riding Toyota speeding along California's Pacific Coast Highway or 'PCH' at Huntington Beach